INTRODUCTION TO
ELECTRONICS
FOR TECHNOLOGISTS

INTRODUCTION TO ELECTRONICS FOR TECHNOLOGISTS

JOHN PAUL HOFFMAN

RESEARCH ENGINEER

CATERPILLAR TRACTOR CO.

AND

ILLINOIS CENTRAL COLLEGE

HOUGHTON MIFFLIN COMPANY   Boston

Dallas   Geneva, Illinois   Hopewell, New Jersey
Palo Alto   London

Dedicated to:

My wife Roxanna, who typed every word of the manuscript— more than once—and to our children, Laura, David, Ruth, Peterjohn, and Nancy, in appreciation and thanks for their thoughtfulness, encouragement, and sacrifices. And to my parents, Louis and May Hoffman.

Printed in the U.S.A.

Library of Congress Catalog Card Number: 77-074381

ISBN 0-395-25115-X

CONTENTS

PREFACE

This book is intended as an introduction to the field of electronics for technologists. Although many of the topics may appear to *teachers* to be less detailed or mathematically rigorous than they might be, first-time *students* in electronics will find the material very challenging. When students have mastered the material in the text they should be able to undertake a more detailed study, which would overwhelm them if approached without a solid grounding in fundamentals.

In general, the mathematical level required is a knowledge of algebra. Occasionally advanced topics such as calculus and Fourier series are used, since the instructor may choose to use some results obtained by mathematics unfamiliar to the student. Therefore no special effort has been made to eliminate the convenient use of mathematics beyond algebra. Perhaps including such topics here will motivate students to learn more about them.

The bipolar junction transistor (BJT) and the field-effect transistor (FET) are considered in separate chapters. Some topics, such as load lines, are thus covered in two places, but there is considerable advantage in repetition of important topics.

The vacuum triode is given reasonable coverage. Although use of vacuum tubes is definitely declining, technologists can reasonably expect to be called upon to apply technical expertise to circuits and instruments using vacuum tubes, and this situation will no doubt be true for the foreseeable future. The coverage is restricted only in that students studying the vacuum triode chapter will sometimes be referred to the analogous section in the FET chapters.

Some topics are covered so that no further study is necessary. The study of diodes, rectifier circuits, and simple passive-component power-supply filter circuits are essentially complete. However, coverage of electrical properties of materials is purposely very limited.

In curricula of limited scope and duration, the coverage of BJTs, FETs, and their associated circuits and functions is adequate, but in curricula with more scope, longer duration, and a higher degree of terminal competence, a more detailed study of common and special electronic circuits and components would be necessary.

Integrated circuits are now being applied in many cases where discrete transistor or tube circuits were previously used. The circuit design function may gradually become a system design function. It would seem to the author that the circuit design function will be needed for years to come, especially for circuits that perform nonstandard functions. Since an understanding of discrete components and circuits is basic to electronics, this introductory text is primarily concerned with them. Chapter 8, however, gives an introduction to the more common types of linear integrated circuits.

As a background for the study of this material, the student should have a good understanding of dc circuit theory. A study of ac circuit theory should precede or be concurrent with the study of this subject. At some schools and in some curricula the book may be used for second-semester freshmen concurrent with an ac circuit-theory course.

This material has been used several times in a beginning course in electronics. The experience gained indicates that the book can be an important part of a very effective plan of study in electronics. Depending upon the results wanted and the time available, a desirable electronics study plan may consist of one, two, or more courses in electronics. This text constitutes the start of an electronics study plan and is therefore the foundation upon which further studies in electronics can build.

I would like to thank the following individuals who reviewed this text in one or more of its developmental stages: Hollis S. Baird, Northeastern University; Robert C. Carter, Southern Technical Institute, Georgia Institute of Technology; Stephen R. Cheshier, Purdue University; Edward Dreisbach, Pennsylvania State University; Larry Flicker, New York City Community College; Robert C. Helgeland, Southeastern Massachusetts University; Gene Larcher, Morrisville Agricultural and Technical College; and Lowell McCaw, Monroe County Community College.

J. P. H.

INTRODUCTION TO
ELECTRONICS
FOR TECHNOLOGISTS

CHAPTER 0
PRELIMINARIES: SOME CONVENTIONS AND DEFINITIONS

There are several conventions that must be defined or explained before we get on with the main part of this book. Some of these items may seem trivial. If our methods and conventions are explained well here, however, our main concentration later in the book can be on considering concepts rather than on wondering what is meant by a particular term or symbol.

0.1 CURRENT

There are two common ways to use the word *current*. To avoid confusion, we will use the word in only one way. Our definitions for electric charge movement per time are as follows:

Conventional current, or **current,** is the movement of positive charges per time. One ampere is one coulomb per second.

Electron flow is the movement of electrons or negative charges per unit time. One ampere is equal to one coulomb per second.

Conventional current and electron flow are equal in magnitude but opposite in sign or direction. If, for example, we have a circuit with a loop current of 1 A clockwise, we can also correctly say that the electron flow is 1 A counterclockwise.

Historically we find that each definition of electric charge flow rate was commonly accepted in certain portions of the electrical and electronics industry. Engineers almost exclusively use the conventional current definition. The electron-flow approach (sometimes called current) has been widely used in technician training programs. Technical high schools, private electronics schools, and military electronics schools are some notable users of the electron-flow method.

In this book we will always understand the word current to mean conventional current. The author feels that this definition will eventually dominate; and since technologists and engineers are members of the same team, they might as well speak the same language.

0.2   SUBSCRIPTING OF VARIABLES

Many times variables will be subscripted, which will help identify each of many similar quantities. If, for example, we have eight different currents in a complex circuit, we can identify each one separately as $I_1, I_2, \ldots, I_7, I_8$.

Letter subscripts find common usage. They may refer to letter-identified points throughout a circuit or to certain elements of a transistor or vacuum tube. Although the names of the elements of transistors and tubes may not mean anything to the reader yet, we will list them here; and when questions arise later about the meaning of a letter symbol, the reader should refer to the listing shown in Figure 0.1.

Element Name	Letter symbol	Schematic symbol	Device name
Plate	P		
Grid	G		Vacuum tube
Cathode	K		
Collector	C		Bipolar
Base	B		junction
Emitter	E		transistor (BJT)
Drain	D		Field-effect
Gate	G		transistor
Source	S		(FET)

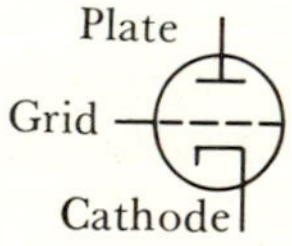

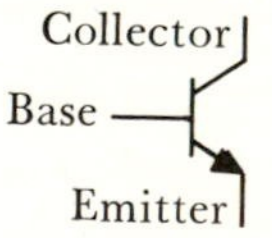

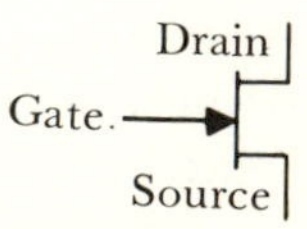

Figure 0.1   Names and symbols for vacuum-tube and transistor devices.

0.3 DOUBLE-SUBSCRIPTED VARIABLES

In some cases more than one subscript is necessary to specifically and clearly identify a variable. In this section we explain how we will be using double-subscripted variables in this book.

A double-subscripted voltage variable is somewhat different from a double-subscripted current variable, owing to the difference in the meaning of the word *voltage* and the word *current*. Current is a measure of the rate of charge movement past a point in a given conductor. By contrast, voltage is the electrical potential difference, or the electromotive force difference between two points or at one point with respect to another.

We define double-subscripted voltage and current variables as follows:

V_{AB} means the voltage at point A with respect to (wrt) point B.

I_{AB} means the current in a conductor connecting points A and B with the positive-defined current direction being from point A to point B.

When speaking about the variables in question, we would say the following: "voltage at A with respect to B" for V_{AB}; and "current from A to B" for I_{AB}.

It is important to learn to say and think of the double-subscripted variables exactly as just quoted. Some thought about double-subscripted variables should convince us that reversing the order of the subscripts will change the algebraic sign of the quantity. In mathematical notation we have

$$V_{AB} = -V_{BA}$$

$$I_{AB} = -I_{BA}$$

Some examples of double-subscripting should be helpful at this point. Figure 0.2 shows a simple circuit with several specific points

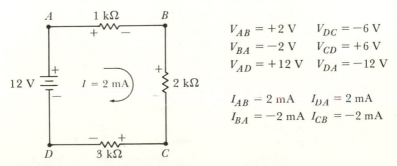

$$V_{AB} = +2\ \text{V} \qquad V_{DC} = -6\ \text{V}$$
$$V_{BA} = -2\ \text{V} \qquad V_{CD} = +6\ \text{V}$$
$$V_{AD} = +12\ \text{V} \qquad V_{DA} = -12\ \text{V}$$

$$I_{AB} = 2\ \text{mA} \qquad I_{DA} = 2\ \text{mA}$$
$$I_{BA} = -2\ \text{mA} \qquad I_{CB} = -2\ \text{mA}$$

Figure 0.2   Example of double-subscripted quantities.

identified by letter symbol. Some of the circuit quantities are listed along with their numerical values. Figure 0.3 shows a circuit containing an insulated-gate field-effect transistor (IGFET). A few of the quantities are given along with their numerical values. Points L and S are in reality the same electrical point, since they are connected with an assumed perfect conducting wire.

If we get lazy and leave off the second subscript, the second subscript is implied. The implied point is a reference, or ground. In Figure 0.3, the symbol V_G means V_{GL} or V_{GS}; it is the dc voltage between gate and ground and/or between gate and source.

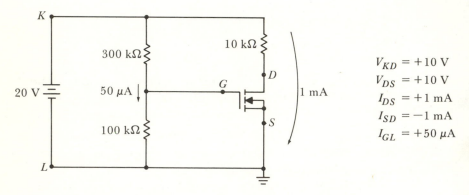

$$V_{KD} = +10 \text{ V}$$
$$V_{DS} = +10 \text{ V}$$
$$I_{DS} = +1 \text{ mA}$$
$$I_{SD} = -1 \text{ mA}$$
$$I_{GL} = +50 \text{ μA}$$

Figure 0.3 Example of double-subscripted variables in a field-effect-transistor (FET) circuit.

0.4 ADDITIONAL SYMBOL AND SUBSCRIPT CONVENTIONS

Letter symbols (for units) and their associated subscripts may be either uppercase or lowercase. Thus there are four possible combinations. Each of the combinations has a separate and distinct meaning. In Figure 0.4(a) we have the general meaning of uppercase and lowercase symbols for units. In (b) we see the general way in which uppercase and lowercase subscripts are used to restrict the wide meaning of the unit symbols. In (c) we have a listing of various unit/subscript combinations and the specific meaning of each. Note that a third subscript is sometimes used. By carefully studying the circuit, the waveshape, and the numerical quantities of Figure 0.5, we should obtain, or verify, our understanding of the symbology used in this book. We believe that this symbology is becoming more or less standard in the electronics industry.

I have found that it is not easy to always use the right combination

Unit Symbol	Meaning
Lowercase (v, i)	Instantaneous value
Uppercase (V, I)	dc, av, or rms value (i.e., steady state)

(a)

Subscript Symbol	Meaning
Lowercase $(a, b,$ etc.$)$	Alternating component value of waveshape
Uppercase $(A, B,$ etc.$)$	Total value of waveshape

(b)

Examples	Meaning
v_{ab}	The instantaneous value of the alternating component of the voltage waveshape at point a wrt point b
i_{ab}	The instantaneous value of the alternating component of the current waveshape from point a to point b
i_{AB}	The instantaneous value of the total current waveshape from point a and point b
V_{ab}	The rms value of the voltage waveshape at point a wrt point b
I_{AB}	The dc value of the total current waveshape from point a to point b
I_{ABM}	The maximum value of the total current waveshape from point a to point b
$V_{AB(AV)}$	The average value of the total voltage waveshape at point a wrt point b

(c)

Figure 0.4   Letter symbol meanings: (a) The general meaning for uppercase and lowercase symbols for units. (b) The general meaning for uppercase and lowercase symbols for subscripts. (c) Specific meaning of several unit symbols with subscripts.

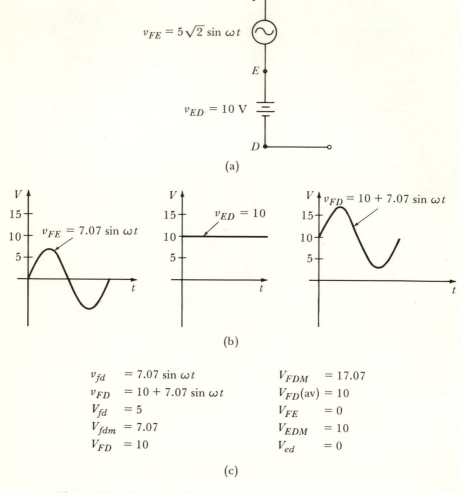

Figure 0.5 Example of symbols and subscripts: (a) circuit connection; (b) waveshapes; (c) subscripted variables with numerical values.

of upper- and lowercase letters for symbols and subscripts when describing a circuit. When concentrating on how a circuit operates, it is easy to get sloppy with symbology; and there are occasions when more than one symbol can be correctly applied.

Presumably, a single reading of this section will not be sufficient to lock all of the symbology conventions permanently in your mind. You may have to refer to it whenever you have a question about the meaning of a symbol. Moreover, many authors are now using what we have called a "standard" symbology in their textbooks, and hopefully nearly all authors will do so in the future. Therefore, you may find this section useful to you many times throughout your career.

0.5   SOME HISTORICAL CONSIDERATIONS

Most of the radios built at the beginning of the radio era were battery-operated. Three batteries were normally required. The one that supplied the energy to heat the filaments of the tubes was called the A battery. The B battery, commonly called the plate supply battery, was needed to supply the voltage necessary for the plates of the tubes. The C battery supplied the necessary bias voltages for the tubes. The words filament, plate, and bias will be defined and used in later chapters.

Some of the terminology of the early days of radio is still with us, even though only a small portion of all electronic equipment is battery-operated or contains vacuum tubes. The letters B and C from the old B battery and C battery are the historical terms of most interest to us. It is still common to use the phrase B+ (read as B plus) or B− (B minus) to indicate the positive or negative side of the dc source used to supply the plate voltage for vacuum tube circuits. C+ and C− have analogous meanings but are less commonly used.

We intend to use the letters B and C in this book in a manner somewhat expanded from their original meanings in order to include the terms in circuits containing junction transistors and field-effect transistors. Thus in this book we will use the B-battery terminology for the voltage supply for the plate, collector, and drain circuits. The plate, collector, and drain are analogous elements in vacuum tubes, bipolar junction transistors (BJT), and field-effect transistors (FET). We will use the C-battery terminology for the bias voltage source in each case (if needed).

Supply voltage sources are often double-subscripted for easy identification. We will follow such a practice in this book. Many times in this book you will see battery symbols labeled V_{BB} and V_{CC}. These represent the supply voltages defined previously. In the actual building of the circuit in question, you may not use a battery to get the necessary dc voltage. We will spend considerable time studying the circuitry to produce the needed dc voltages without a battery. Even though the dc voltages do not always come from a battery, it is convenient to use a battery symbol to represent the presence of a dc voltage.

The practice of using *BB* and/or *CC* as subscripts for the dc voltage supply sources for all devices is not common. However, we have chosen this method for what we believe are good and valid reasons. We will present the reasons shortly.

Many authors use a different voltage-source subscript letter for each distinct type of device under discussion. It is common to see the subscript *CC* for the collector supply of bipolar transistors, *DD* and *SS* for the drain and substrate supplies for field-effect transistors, and *BB* and *CC* for the plate and bias supplies for vacuum tubes. I have no quarrel with this approach as long as a single type of device is under discussion.

As the state of the art in electronics advances, we see more and more circuits, systems, and integrated devices that include more than one type of device. Often these multiple-device-type circuits use only one dc supply voltage. Thus, finding the appropriate subscript to use presents a problem. The problem is solved, however, with the system that we have chosen; or with any other system that uses a subscripting scheme independent of the type of device used.

Another historical term that needs some comment is ground. In the strictly and absolutely correct sense, the term *ground* refers to the electric potential (voltage) of the earth. If a circuit diagram has a ground (⏚) symbol in it, then that point must be connected to earth through a conducting path (wire) if the strict meaning of the ground symbol is being followed.

In addition to the strict meaning of the term *ground,* there is a more generalized and more widely used meaning. In the generalized concept, the ground symbol merely identifies a reference point or a common point within the circuit. The reference point identified by a ground symbol may or may not be electrically connected to the earth. The generalized meaning is used in this text.

0.6   APPLICATION OF KIRCHHOFF'S VOLTAGE LAW

It has been the author's observation that there is a lack of consistency in the practical application of Kirchhoff's voltage law (KVL). Our purpose here is to show you the author's way of applying the law so that there will be no misunderstanding when you read some of the equations derived by application of Kirchhoff's voltage law.

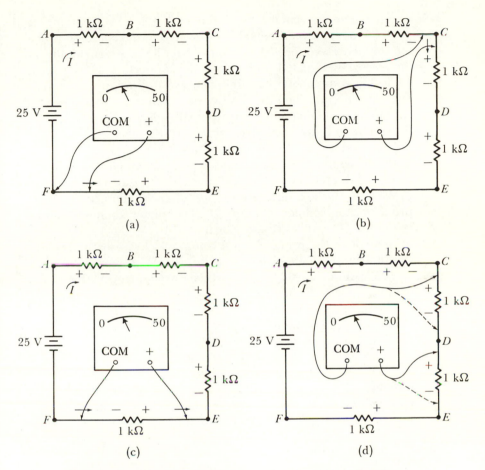

Figure 0.6 Circuit used to explain the application of KVL.

Many times in the study of electricity and/or electronics we must solve equations, either singly or simultaneously, in order to find some unknown quantity (or quantities) of interest. Kirchhoff's laws provide a convenient method of writing the necessary equations. It seems desirable to write KVL equations exactly as we would take experimental data in the laboratory on the same circuit. The methods used in this book do in fact use the same procedures and thought processes for writing KVL equations as those used for taking experimental data.

Consider the circuit of Figure 0.6(a), where we have a voltmeter pictorially shown inside the schematic diagram of an electric circuit.

Assume that we actually build this circuit in the laboratory and proceed to take experimental data. Note that the common probe of the voltmeter is connected to point F. Now let us systematically move the positive-designated probe along the connecting wires in a counterclockwise direction. The first nonzero indication we get on the voltmeter occurs when the positive-designated probe makes contact with point E. At that time, the meter will indicate the voltage at point E with respect to F (that is, at E wrt F), which is symbolically represented as V_{EF}. In this case, $V_{EF} = +5$ V.

As we continue to systematically move the positive-designated probe counterclockwise around the circuit, leaving the common probe at F, the next different voltage we encounter occurs when the probe makes contact at point D. The meter then indicates the voltage at D wrt F (V_{DF}), which is $+10$ V. Further systematic movement of the probe along a counterclockwise path around the circuit will yield, in order, $V_{CF} = +15$ V, $V_{BF} = +20$ V, $V_{AF} = +25$ V, and $V_{FF} = 0$ V. The last measured quantity, that is, $V_{FF} = 0$ V, is an experimental validation of Kirchhoff's voltage law.

The starting point and the path direction are unimportant in using the KVL equation. To apply the same method with a different starting point and a different path direction, let us refer to the situation shown in Figure 0.6(b). Point C is the reference point and the common probe is connected to this reference point. As we systematically move the positive-designated probe in a clockwise path around the circuit, the meter indicates in sequence the voltages $V_{DC} = -5$ V, $V_{EC} = -10$ V, $V_{FC} = -15$ V, $V_{AC} = +10$ V, $V_{BC} = +5$ V, and $V_{CC} = 0$ V.

When using the KVL equation we are usually interested in each individual voltage drop and voltage rise around a loop. This may be accomplished with only one change from the sample measurements we have already performed on the circuit shown in Figure 0.6(a) and (b). The change involves moving the common probe (that is, the reference point) after each measurement, so that the voltmeter indicates the voltage across a single component.

Let us now experimentally take data around a closed loop so that the resultant individual voltage values can be directly and immediately substituted into the KVL equation that we would get for the same path. The diagram for the circuit we will be considering is shown in Figure 0.6(c). The starting point, and the reference for the first voltage measurement, is point F. We will traverse a counterclockwise path. For the connection shown, the voltmeter will indicate $V_{EF} = +5$ V. If we move each probe counterclockwise by one component increment (as indicated by the direction-of-motion arrows across the probes), the voltmeter will indicate $V_{DE} = +5$ V. Continuing around the entire loop, we obtain $V_{CD} = +5$ V, $V_{BC} = +5$ V, $V_{AB} = +5$ V, and $V_{FA} = -25$ V.

If we write the KVL equation for the circuit of Figure 0.6(c) in exactly the way that we have just done our experimental measurements, we will certainly have a logical and consistent method. Starting at point F and going counterclockwise (as in measurements), we obtain the following:

$$V_{EF} + V_{DE} + V_{CD} + V_{BC} + V_{AB} + V_{FA} = 0$$

Since we have written the equation in the same manner as we performed our experimental measurements, the double subscripts are in the same order for the equation and for the experimental data. Repeating the equation and the experimental data exactly in the order performed, we have

$$V_{EF} + V_{DE} + V_{CD} + V_{BC} + V_{AB} + V_{FA} = 0$$

and

$$+5 \text{ V} + 5 \text{ V} + 5 \text{ V} + 5 \text{ V} + 5 \text{ V} - 25 \text{ V} = 0$$

In Figure 0.6(d) we choose to write the KVL equation and then take experimental data starting at point C and traversing a clockwise path. The first voltmeter connection is shown with probes indicated as solid lines, and in the second position the probes are indicated with dashed lines. The results of our efforts, with both equation and experimental data, will be

$$V_{DC} + V_{ED} + V_{FE} + V_{AF} + V_{BA} + V_{CB} = 0$$

and

$$-5 \text{ V} - 5 \text{ V} - 5 \text{ V} + 25 \text{ V} - 5 \text{ V} - 5 \text{ V} = 0$$

Stated in an unconventional way, the KVL equation would be "the algebraic sum of the voltages around any path from point A to point A is zero." Stated mathematically we have

$$V_{AA} \text{ (any path)} = 0$$

The repeated subscript, that is, the same starting and ending point, is another way of specifying a closed path as in the conventional way of stating KVL.

We have stated that our method of writing KVL equations is the same as that used in experimental measurements. Since we can experimentally measure the voltage between any two points as the algebraic sum of voltage drops and voltage rises between the two points of interest, we should also be able to use the same technique to write the appropriate equation. In other words, we should be able to find, for example, V_{XY}, either experimentally or symbolically. Using the same mathematical terminology as for KVL, we will be finding

V_{XY} (any path)

As an example, let us determine the voltage between points B and D. Specifically we want to know the voltage V_{DB}, the voltage at D wrt B for the circuit of Figure 0.6(c). Since point B is the reference, we place the common-designated voltmeter probe at point B. Then we note that there are two circuit paths connecting points B and D. Choosing the clockwise path and moving the voltmeter probes in order to measure the voltage across each component in the clockwise path between the reference (point B) and point D, we obtain

$V_{DB} = -5 \text{ V} - 5 \text{ V} = -10 \text{ V}$

If we use the same technique to write the symbolic voltage between the same points, the result is

$V_{DB} = V_{CB} + V_{DC}$

Finding the voltage V_{DB} by taking the counterclockwise path between B and D, we get the experimental result

$V_{DB} = +5 \text{ V} - 25 \text{ V} + 5 \text{ V} + 5 \text{ V} = -10 \text{ V}$

and the symbolic result

$V_{DB} = V_{AB} + V_{FA} + V_{EF} + V_{DE}$

Note that we get the same numerical result regardless of the path taken. The symbolic results are also valid and equivalent for each path.

By now you should be completely familiar with, and understand, the method of writing loop equations used in this book. The desirable features of our methods are as listed below.

1. There is only one method to know or remember. It applies equally to equations in which voltages are represented symbolically and to experimental voltage measurements.

2. The reference point is considered as in other common experiences, that is, the reference is the observer's position. When experimentally measuring voltages, the measurer imagines a position at the reference point of the measurement, that is, the common probe of the voltmeter. The indicated voltage, then, is the voltage felt, or sensed, at the other probe with respect to the measurer's position (reference).

3. By emphasizing the reference, and measurements with respect to a reference, we minimize the probability of students' talking of the false concept of, for example, voltage "going from A to B."

The method we have described, which will be used in this text, is not the only method available. Any correct method will yield the same results that we have obtained and will obtain throughout the book. With another method, however, the results may appear to be different, but probably only with regard to something as simple as reverse-order subscripts. In any case, any two correct methods will yield results that can be shown to be equivalent with only minor algebraic manipulation.

0.7 SCHEMATIC SYMBOLS

Anyone studying electronics will soon discover that in certain areas there is a lack of standardization of schematic symbols. Our purpose in this section is to inform you of what to expect in this book.

Figure 0.7 shows the schematic symbols that will be used to represent electrical sources in this book. Independent sources are represented by circles and dependent sources are represented by diamonds. An exception is sometimes made in the case of an independent dc voltage source, for which a battery symbol may be used.

While circles and diamonds are used to indicate independent or dependent sources, respectively, we need a method of differentiating between voltage sources and current sources. A current source is

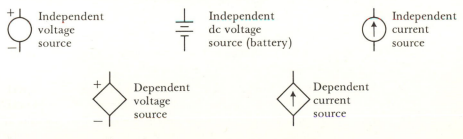

Figure 0.7   Schematic symbols for electrical sources.

identified by an arrow showing the direction of positively defined conventional current. A voltage source will have associated polarity signs to show the positively defined voltage polarity reference. In the case of time-varying sources, the waveshape produced by the source may be drawn within the symbol or nearby.

The subject of schematic symbols also includes the symbols to be used for electronic devices such as transistors, vacuum tubes, diodes, and so forth. In this book we have tried to use commonly accepted symbols, even though in some cases no standardization is apparent. Rather than list here all the symbols chosen, each will be presented with the appropriate subject matter.

EXERCISES

QUESTIONS

Q0.1 What is the difference between the quantities known as conventional current and electron flow?

Q0.2 Double subscripts may be used with either voltage or current quantities. What is the exact meaning of the subscripts in each case?

Q0.3 What are the various implications of the word ground and/or the use of the ground symbol ($\doteq$)?

Q0.4 What were the original applications of the A, B, and C batteries?

PROBLEMS

P0.1 Find the magnitude and direction of the conventional current in each branch of the circuit of Figure 0.8.

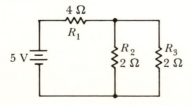

Figure 0.8   See Problem 0.1.

P0.2 Solve for the numerical value (including polarity) of each of the following quantities, as defined in the circuit of Figure 0.9:

$I_{AB}, I_{AC}, I_{BC}, I_{CD}, I_{BE}, I_{ED}, I_{CA}, I_{DC}, I_{EB}, V_{AE}$

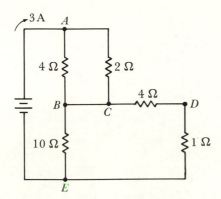

Figure 0.9   See Problem 0.2.

P0.3 Find numerical values for the following listed quantities for the waveform of Figure 0.10.

$V_{AB}, V_{ab}, v_{AB}, v_{ab}$

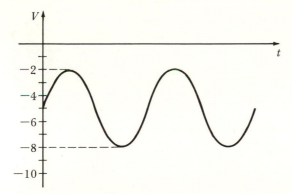

Figure 0.10   See Problem 0.3.

P0.4 Write an equation for V_{FB} using double-subscripted voltage variables as you take the clockwise path from B to F. The circuit is shown in Figure 0.11.

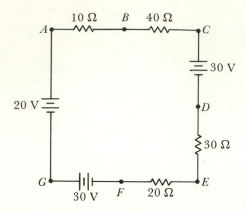

Figure 0.11   See Problem 0.4.

P0.5 Repeat Problem 0.4 using a counterclockwise path.

P0.6 Repeat Problems 0.4 and 0.5 using the appropriate numerical quantities for the previous double-subscripted variables.

CHAPTER 1
SEMICONDUCTING MATERIALS

Coverage of the atomic structure of materials is purposely very brief. The purpose of this chapter is to review the Bohr model of an atom and to review some of the basic concepts and terminology associated with the atomic structure of materials. Silicon and germanium are widely used in electronic components, so we will be primarily interested in these materials.

Pure silicon and pure germanium are seldom used in electronic components. Since this is the case, we will talk of adding impurities to silicon or germanium. Adding impurities (doping) leads to the formation of *PN* junctions and many electronically useful devices.

1.1 BOHR MODEL OF AN ATOM

The Bohr model of an atom of any element is made up of a nucleus and orbiting electrons. The nucleus contains protons and neutrons. The protons are particles of positive charge, and the neutrons are electrically neutral. The orbiting electrons are said to be contained in shells. The shells are assigned letters for identification, with the K shell physically closest to the nucleus and the Q shell the greatest distance from the nucleus. The Bohr atom is shown in Figure 1.1.

There are definite limits to the number of electrons that may occupy any shell. For example, the maximum number of electrons that may be contained in the K shell is 2. The L shell may contain as

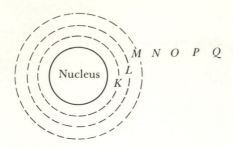

Figure 1.1   Bohr model of atom.

many as 8, the M shell may contain 18, and the N shell may contain 32.

The physical and chemical properties of elements are determined by the number of electrons, protons, and neutrons in the makeup of their atomic structure. Generally, a stable configuration occurs when there are eight electrons in the outer shell (two if the K shell is the only shell). The electrons in the outer shell are called *valence electrons.* The valence electrons are important in determining the properties of the element.

We note from Table 1.1 that both silicon and germanium atoms have four electrons in the outer shell. Recall also that eight electrons in the outer shell results in a stable configuration. When silicon (or germanium) is in the solid state, the individual atoms form a definite crystalline structure. The crystalline structure is such that each atom shares a valence (outer shell) electron with four neighboring atoms. Since each atom in the crystal structure now has the effective use of eight valence electrons, a stable structure results. There is an attractive force among the atoms sharing their valence electrons. This attractive force is the bond that holds the crystal structure together in a definite pattern. This particular type of bonding is called *covalent bonding* because it is the sharing of the valence electrons between atoms that causes the attractive bonding forces. Figure 1.2 shows, in a two-dimensional sketch, the electron sharing and cova-

Table **1.1**   A listing of the electron content of each shell of the elements silicon and germanium

| Element | Name of shell | | | |
	K	L	M	N
Silicon	2	8	4	
Germanium	2	8	18	4

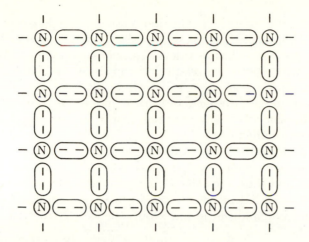

Figure 1.2 Covalent bonding. The electrons within the oval loops are the covalently bonded electrons.

lent bonding that occurs in either pure silicon or pure germanium crystals. The circles labeled N represent the nucleus and all electrons except the four valence electrons. The valence electrons are shown as small dashes. The four valence electrons are shown close to their respective nuclei. Actually, a shared electron spends an equal amount of time with each associated nucleus and on the average is equidistant from each associated nucleus. The two-dimensional view of the germanium or silicon crystal showing the crystalline structure and the covalent bonding is a simplification of the actual three-dimensional phenomena.

1.2   INSULATORS, CONDUCTORS, AND SEMICONDUCTORS

Materials are classified as insulators, conductors, or semiconductors according to how freely they conduct electric current. A material is a conductor if the ratio of current to voltage is "high." It is an insulator if the ratio of current to voltage is "low." Thus a good conductor is a poor insulator, and vice versa. Generally, the atomic structure of a good conductor contains one or two loosely held electrons in the outer orbit (shell). These loosely held electrons are relatively free to move throughout the material. A charge movement, of course, constitutes a current flow. Gold, silver, and copper each have one electron in the outer shell and each is also a very good electrical conductor. Insulators usually have nearly eight electrons, which is a stable state, in the outer shell. The inert gases have eight electrons in the outer shell and are insulators unless ionized by high electric forces (voltage). Sulfur has six valence electrons and iodine has seven valence electrons; both are relatively good insulators. There is no well-defined dividing line that separates insulating materials from conducting materials. In fact, our main interest in the electrical properties of materials is with materials that are neither good insulators nor good conductors.

Semiconductor is the term applied to materials that are located in a broad transition region between those materials which we classify as insulators and those we classify as conductors. Germanium and silicon are considered to be semiconductors. As mentioned previously, germanium and silicon have four valence electrons each.

1.3   *N*-TYPE AND *P*-TYPE SEMICONDUCTING MATERIALS

Pure germanium or pure silicon forms a monocrystalline structure as shown in Figure 1.3(a). If a trivalent atom, that is, an atom with three valence electrons, is included in the crystal structure, the result

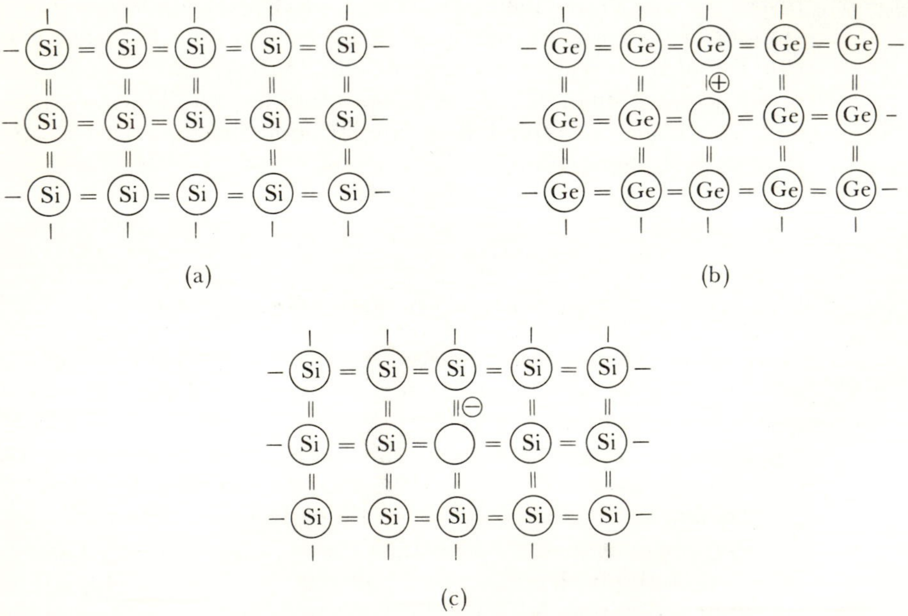

(a)

(b)

(c)

Figure 1.3   Electronic configuration in crystal structure: (a) pure silicon; (b) *P*-type germanium; (c) *N*-type silicon.

is as shown in Figure 1.3(b). The trivalent atom is called an impurity atom. The impurity atom is electrically neutral since it has the same number of positive charges (protons) in the nucleus as it has orbiting electrons. Even though the material as a whole is electrically neutral, we see that one covalent bond cannot be made. The circled plus sign in Figure 1.3(b) indicates the absence of an electron in a desired covalent bond. This absence is called a *hole,* which is a positive charge with respect to valence electrons and covalent bonds. Remember, however, that the material is electrically neutral as a unit. It contains as many negative charges as positive charges.

When a pentavalent impurity atom is incorporated in a germanium or silicon crystal, that is, an atom with five valence electrons, the resultant electronic structure is as shown in Figure 1.3(c). Every desired covalent bond is made and, in addition, an unbound electron is present. The unbound electron is shown as a circled minus sign. It is in excess for covalent bonding; but each atom, including the impurity atom, is electrically neutral. The entire crystal is also electrically neutral.

A germanium or silicon crystal that has added impurity atoms is said to be doped. Although doping levels vary widely, the number of doping or impurity atoms is small in percentage. Doping may be spoken of as so many parts of impurity atoms per million germanium or silicon atoms. When trivalent atoms are used in doping, the resultant crystal is called *P-type*. The *P* is for positive, since one negative charge (electron) is absent in a desired covalent bond. The absence of a negative charge is equivalent to the excess of one positive charge, or hole. When pentavalent atoms are used in the doping process, excess negative charges (electrons) are present in terms of covalent bonding. Thus, pentavalent doped crystals are called *N-type*.

Doped germanium or silicon has higher conductivity than does the pure germanium or silicon. The conductivity is directly related to the concentration of impurity atoms introduced in the doping process. It is understandable that conductivity increases with doping, since in *N*-type material, for instance, there are non-covalently bonded electrons (one for each impurity atom) that are loosely held and thus easily available for current conduction. In *P*-type materials, the holes are loosely held and are available for current conduction.

The movement of holes in *P*-type material seems rather artificial. The concept is commonly used, however, so we should be aware of the process. Figure 1.4 is a two-dimensional view of the electronic structure of *P*-type semiconducting material. The trivalent impurity atom leaves one covalent bond unfilled. This absence is called a hole. If a nearby covalently bonded electron is by any means induced to

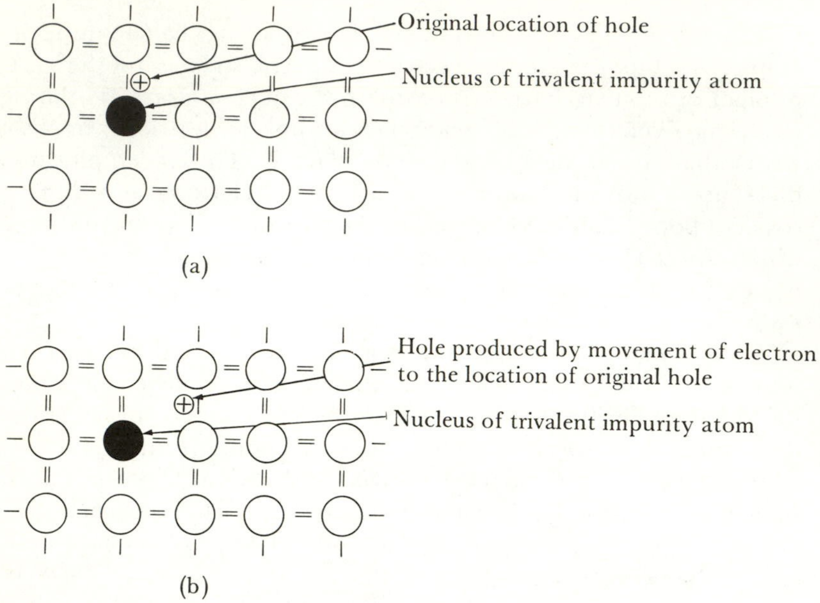

Original location of hole

Nucleus of trivalent impurity atom

(a)

Hole produced by movement of electron to the location of original hole

Nucleus of trivalent impurity atom

(b)

Figure 1.4   Electronic structure of *P*-type semiconductor (a) before and (b) after movement of a hole.

leave its present location and fill the hole, a new hole will appear at the location from which the electron departed. In Figure 1.4(a) we see a hole in its original location associated with its own nucleus. In Figure 1.4(b), an electron has moved to the left to fill the original hole and a new hole has appeared at a location to the right of the original hole. We thus talk of the hole as having moved to the right. The impurity atom at the location of the original hole now has a net negative charge. The neighboring atom that gave up an electron now has a net positive charge (hole). The entire crystal is electrically neutral, however, because no charge has been added to or taken from the crystal.

1.4   THE *PN* JUNCTION

A *PN* junction is represented as in Figure 1.5. The crystalline structure must be continuous through the junction or transition region between the *P* and *N* parts of the crystal structure. The *P*-type material is represented with plus (+) signs to indicate the presence of unbound holes; and the *N*-type material has minus (−) signs to indicate the presence of unbound electrons.

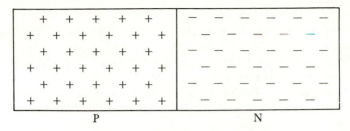

Figure 1.5 Semiconductor *PN* junction.

   The entire crystal in Figure 1.5 is electrically neutral, since each individual atom, including impurity atoms, is neutral. However, owing to thermal agitation, there will be some movement of charges throughout the material that will cause part of the crystal to be negatively charged and part to be positively charged even though the entire crystal is electrically neutral. Any electron in the *N*-type material near the junction that "acquires" enough energy to break the weak attractive forces holding it will more than likely cross the junction to fill a hole in the *P*-type material. As a result of this one electron moving across the junction, the *P*-type material is now negative and the *N*-type material is positive. A voltage potential develops across the junction as electric charges continue to cross the junction. Eventually an equilibrium is reached in which the voltage across the junction is great enough to limit the number of charges that cross the junction. At equilibrium a depletion region is formed at the junction that is void (depleted) of loosely held charge carriers, as shown in Figure 1.6. The circled minus signs indicate electrons

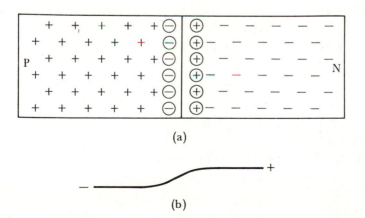

(a)

(b)

Figure 1.6 *PN* junction at equilibrium: (a) electron configuration; (b) potential hill.

that have crossed the junction and have filled a hole and are now bound in a covalent bond at that location. The circled plus signs are holes formed by the departure of electrons across the border. The charges that have crossed the junction cause a *potential hill* to be formed, which is pictured in Figure 1.6(b).

A junction of *P*-type and *N*-type semiconductor materials has useful properties. These properties are discussed and put to use in the next chapter.

EXERCISES

QUESTIONS

Q1.1 What name is given to the electrons in the outer shell of an atom?

Q1.2 What bonding forces cause the elements germanium and silicon to form a crystalline structure?

Q1.3 What are the differences between a conducting material, a semiconducting material, and an insulating material?

Q1.4 Is there any relation between the number of electrons in the outer shell of an atom and the resistivity of the element? What?

Q1.5 What is meant by a trivalent atom or a pentavalent atom?

Q1.6 What is the difference between *N*-type and *P*-type semiconductor materials in terms of the doping materials used?

Q1.7 What is a hole in a semiconducting material?

Q1.8 How does the conductivity of a semiconducting material vary with doping concentration?

Q1.9 What is a *PN* junction?

Q1.10 If I had a cube of *N*-type semiconductor and a cube of *P*-type semiconductor, could I make a *PN* junction? Why or how?

Q1.11 What is, and what causes, a potential hill?

Q1.12 What causes a depletion region to be formed at a *PN* junction?

Q1.13 What are the various components that make up the Bohr model of an atom? Discuss each.

CHAPTER 2
TWO-ELEMENT ELECTRONIC DEVICES

In this chapter we intend to familiarize the reader with an electronic component called a *PN* junction diode. We will discuss its useful characteristics and then analyze simple circuits containing diodes.

2.1 THE *PN* JUNCTION DIODE

When conducting leads are attached to a *PN* junction, as shown in Figure 2.1(a), the device has useful characteristics and is called a *PN junction diode.*

If a battery is applied to a *PN* junction diode, as shown in Figure 2.1(b), we find that the loosely bound electrons in the *N*-type material are attracted toward the positive battery terminal and away from the junction. Similarly, the holes in the *P*-type material are attracted toward the negative battery terminal and away from the junction. The main effect of applying the battery in this direction has been to widen the depletion region. Since all loosely held charges tend to move away from the junction, we have no charge carriers crossing the junction and no current. Applying a voltage to a *PN* junction in this polarity is called *reverse bias.* The battery merely adds to or reinforces the potential hill.

A forward-biased *PN* junction is shown in Figure 2.1(c). Loosely held electrons in the *N*-type region are repelled or driven away from the negative terminal of the battery toward the junction. The holes

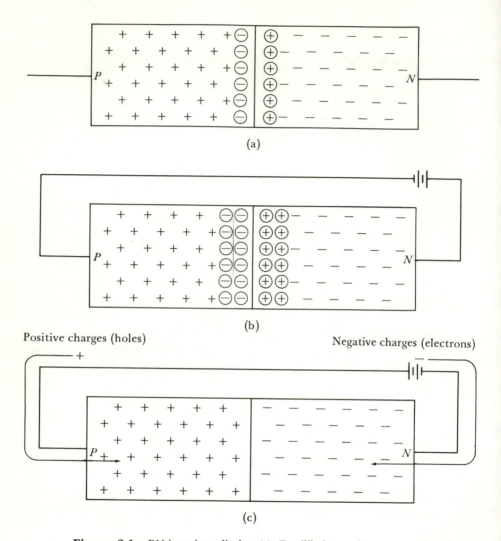

Figure 2.1 *PN* junction diode. (a) Equilibrium; (b) reverse-biased; (c) forward-biased.

in the *P*-type region are repelled away from the positive terminal of the battery toward the junction. At the junction, the positive holes and the negative electrons combine. [The combination of an electron and the absence of an electron (hole) is nothing. An electron and a hole cancel into nothingness.] Thus the drift of charges toward the junction continues and we have a current. The circuit shown in Figure 2.1(c) has conventional current, that is, a movement of positive charges, in a counterclockwise direction. Electron flow, a movement of electrons, is in the clockwise direction. (Refer to current convention discussed in Chapter 0.)

2.2 JUNCTION-DIODE CHARACTERISTICS

Our discussion up to the present time has been concerned with the internal microscopic physical characteristics of the semiconductors germanium and silicon. In particular, we have been interested in doped semiconductors and the *PN* junction. The coverage has been brief and incomplete. It is hoped, however, that the reader has achieved some understanding that will be helpful in the study of solid state devices and circuits.

Some acquaintance with microscopic quantities is desirable. As practicing technologists concerned with the design, operation, or maintenance of electronic devices and circuits, however, it is considerably more important to have a thorough understanding of the external terminal characteristics of solid state devices as well as high vacuum and gaseous devices.

At this point, we will talk about devices in terms of the voltage and/ or current characteristics at the terminals. This is an excellent time to review the letter symbols and subscripts we will be using. The symbols were explained in Chapter 0.

In our discussion of the internal microscopic operation of a *PN* junction, we saw that a *PN* junction will permit current flow in one direction but none (or very little) in the other direction. If we assume that the diode has zero resistance in the conducting (or forward) direction and infinite resistance in the nonconducting (or reverse) direction, the *I-V* terminal characteristic will be as shown in Figure 2.2(b). This is called an ideal characteristic, or the *I-V* characteristic of an ideal diode. The *I-V* characteristic of a physically realizable junction diode, which may be readily purchased, is as shown in Figure 2.2(c). The schematic symbol for a solid state diode is shown in Figure 2.2(a).

The ideal diode characteristic shown in Figure 2.2(b) has never been realized. When making circuit calculations, however, many times we assume a diode to be ideal even though we are fully aware that it is not ideal. The reason for assuming the ideal case is to make the calculations more manageable. The actual characteristic of a diode is nonlinear and cannot easily be handled with the ordinary mathematics with which we are familiar.

In some cases the error introduced into the solution of a circuit problem by assuming an ideal diode characteristic is very small. In

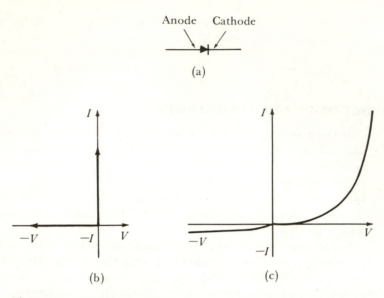

Figure 2.2 *PN* junction diode: (a) symbol; (b) idealized *I-V* characteristic; (c) typical *I-V* characteristic.

such a case it is completely proper to make such an assumption because of the increased ease of computation. There are cases, however, in which either the error is large or where even a small error cannot be tolerated.

2.3 DIODE EQUIVALENT CIRCUITS

In cases where we need more accuracy than can be obtained by using the simple ideal diode approximation, we generally make approximations using only linear elements to represent the diode. We use the nonlinear actual characteristic only as a last resort, which will not be covered here.

Figure 2.3 shows three commonly used diode equivalents. Figure 2.3(b) shows the simplest nonideal equivalent. The circuit consists of an ideal diode in series with a linear resistance. The *I-V* characteristic consists of two straight-line segments that join at the origin. Figure 2.3(c) shows a more accurate equivalent circuit that consists of an ideal diode, a resistance, and a battery. The *I-V* characteristic for this equivalent circuit also consists of two straight-line segments. The

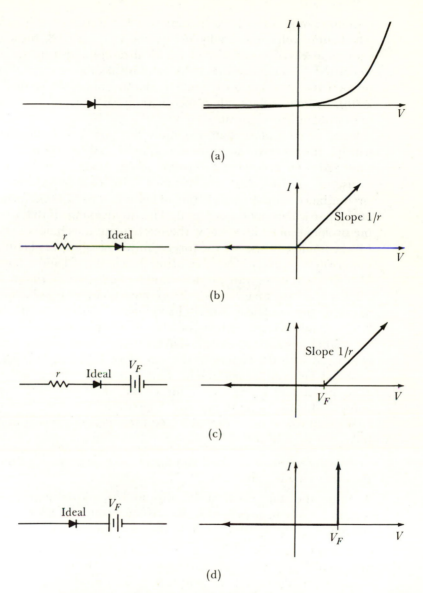

Figure 2.3 *PN* junction-diode equivalent circuits and *I-V* characteristics: (a) actual characteristic; (b), (c), and (d) approximate characteristic.

common point of the line segments is at $I = 0$, V = battery voltage. The battery voltage is usually less than 1 V. Figure 2.3(d) shows an equivalent circuit consisting of an ideal diode and a battery. The associated I-V characteristic is like the ideal diode except that the turn-on voltage is not zero. The I-V characteristics shown in Figure 2.3(b), (c), and (d) are called piecewise-linear because they contain only straight-line segments.

Since we have more than one equivalent circuit available, each of which is the approximate representation of a PN junction diode, we will need some criteria to determine which diode equivalent circuit to use in any particular circumstance. The least error, under any circumstance, would result if the actual nonlinear I-V characteristic of the particular diode were used. The disadvantage is that to do so the problem solver must know the mathematical techniques needed for the nonlinear case. Of the three equivalents shown in Figure 2.3, the greatest accuracy is possible using the circuit of Figure 2.3(c) (an ideal diode in series with a resistor and battery). We use the circuit of Figure 2.3(c) any time we want the greatest accuracy possible without going to the nonlinear actual characteristic. The circuit of Figure 2.3(b) could be used, with reasonably small error, in circuits where the signal voltages are large. Large in this case means large as compared with the battery in the circuit of Figure 2.3(c), which is probably between 0.2 and 0.8 V. The circuit of Figure 2.3(d) is best used in circuits where the other circuit resistances are large as compared with the resistance r of Figure 2.3(c).

We summarize our coverage of equivalent-circuit usage with the following rules of thumb.

1. Use the circuit of Figure 2.3(c) for greatest accuracy regardless of the other circuit components.

2. When the other circuit resistances are large, as compared with r, we can usually neglect r, as in the circuit of Figure 2.3(d), without introducing very much error into the calculated results.

3. When the signal voltages are large, as compared with the battery voltage of Figure 2.3(c), we can usually neglect the battery voltage, as does the circuit of Figure 2.3(b), without introducing very much error into the calculated results.

4. When the circuit resistances are large, as compared with r, and in addition the signal voltages are large, as compared with the battery voltage, it is oftentimes satisfactory to use the ideal diode as being equivalent to the actual diode. The ideal diode representation is also used, even when the conditions above are not met, when it seems desirable to have a quick and easy answer to a particular problem. It must be remembered, however, that the calculated data will be ballpark rather than extremely accurate.

5. Rules 2, 3, and 4 can perhaps be summarized as follows:

(a) If the equivalent resistance r is small and seemingly insignificant in comparison with other circuit resistances, delete resistance r from the equivalent circuit.

(b) If the equivalent circuit battery voltage is small and seemingly insignificant in comparison with the signal voltage levels, delete the battery voltage from the equivalent circuit.

2.4 EQUIVALENT-CIRCUIT COMPONENT VALUES

We have discussed several diode equivalent circuits and given some general rules about which equivalent circuit is best applied to a given situation. The next step, then, is to determine the numerical values of resistance r and the diode forward-conducting voltage V_F. The forward-conducting voltage is sometimes called the *turn-on voltage*.

The *I-V* characteristic shown in Figure 2.4 represents a diode for which we want to find the equivalent circuit. Determination of the numerical quantities is essentially a graphical process. A straight-line approximation is graphically superimposed on the actual *I-V* characteristic in Figure 2.4(a). The straight-line approximation shown is not unique. Each person making such a construction would probably obtain a slightly different result. For the straight-line approximation shown, the turn-on voltage is 0.6 V. Thus V_F is 0.6 V. The value of the dynamic resistance r is determined by the slope of the straight-line equivalent. The quantities ΔV and ΔI are obtained from a graphical construction as shown. In general, resistance is the ratio of voltage to current. Specifically the dynamic resistance r is

$$r = \frac{\Delta V}{\Delta I}$$

The functional relationship between slope and dynamic resistance is

$$\text{Slope} = \frac{\text{rise}}{\text{run}} = \frac{\Delta I}{\Delta V} = \frac{1}{r}$$

In this specific example,

$$r = \frac{\Delta V}{\Delta I} = \frac{80 \text{ mV}}{4 \text{ mA}} = 20 \text{ } \Omega$$

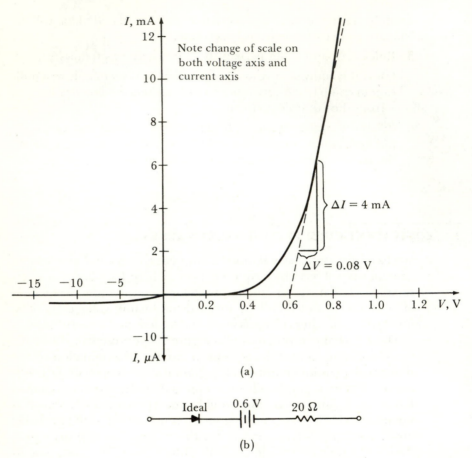

Note change of scale on both voltage axis and current axis

$\Delta I = 4$ mA

$\Delta V = 0.08$ V

(a)

Ideal   0.6 V   20 Ω

(b)

Figure 2.4 *PN* junction diode: (a) characteristic curve; (b) equivalent circuit.

The total equivalent circuit for the diode whose characteristic is shown in Figure 2.4(a) is shown in Figure 2.4(b).

It is quite possible that for a different application, another set of equivalent-circuit values would be more appropriate. If, for instance, we knew that in a certain application the current through the diode in the forward direction, when turned on, would be in the range of 1 to 4 mA, the equivalent circuit derived from the construction shown in Figure 2.5 would be quite accurate—much more accurate than the equivalent circuit we derived from the construction shown in Figure 2.4.

The numerical values are as follows:

$V_F = 0.43$ V by inspection

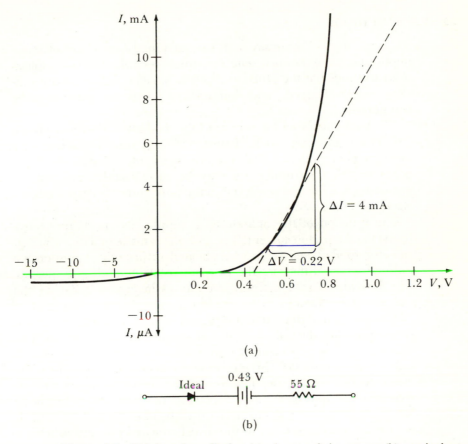

Figure 2.5 *PN* junction diode: (a) characteristic curve; (b) equivalent circuit.

$$r = \frac{\Delta V}{\Delta I} = \frac{0.22 \text{ V}}{4 \text{ mA}} = 55 \text{ }\Omega$$

We have now derived two different sets of equivalent circuit values from the same diode *I-V* characteristic curve. We should now be in a position to logically conclude that whether we describe the diode equivalently as 0.6 V and 20 Ω or as 0.43 V and 55 Ω, or some other consistent set of parameters, is dependent on the application and/or circuit involved. If the accuracy demands it, we will have to make an appropriate choice. In many cases, however, either derived set of values would be satisfactory. In addition, we will find that in many cases we can use one of the simpler equivalent circuits, including the ideal diode alone.

2.5   VACUUM DIODE

Although it was historically first, the vacuum diode is now of lesser importance than the solid state *PN* junction diode. In most applications we find that the junction diode is smaller, lighter, more efficient, more rugged, and longer-lived than its vacuum tube counterpart.

The construction of the vacuum tube diode is as shown in Figure 2.6. The entire unit is sealed under high-vacuum conditions in a metal or glass container. The container is usually called an envelope. Some of the structural parts may be metals and some may be insulators, but the end result is that the parts must be supported and electrically insulated.

The function of the cathode is to supply electric charges (electrons) to take part in current. The cathode is made of a material that will emit (boil off) electrons when heated sufficiently. Tungsten is a cathode material that has the proper electron-emitting properties. The cathode may be a separate entity as in Figure 2.6(a), or the cathode and filament may be combined as in Figure 2.6(b).

The function of the filament is merely to supply the high temperatures needed by the cathode material in order to emit electrons. The filament as shown in Figure 2.6(a) is separate and insulated electrically from the cathode. This type of construction is referred to as the indirectly heated cathode. The directly heated cathode type of construction is shown in Figure 2.6(b). The filament and the cathode functions are combined and performed by one physical element. It is probably easiest to think of the directly heated cathode as being a filament made of (or coated with) an emitting material.

The plate of a vacuum tube is the element to which the emitted electrons are attracted if and when a positive voltage is applied to the plate with respect to the cathode. Anode is another term for plate.

2.6   VACUUM DIODE *I-V* CHARACTERISTICS

The operation of a vacuum tube diode is rather easily explained. The electrons emitted or boiled off the cathode form what is called an electron cloud or space charge in the vicinity of the cathode. If and when the plate voltage is positive with respect to the cathode, some of the space-charge electrons will be attracted to the plate. This produces a current through the device. If the plate is negative with respect to the cathode, the space-charge electrons will definitely not be attracted to the plate and thus there will be no current through

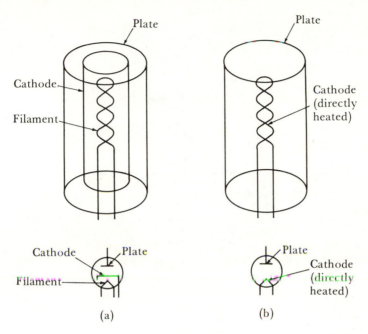

Figure 2.6 Vacuum-diode construction and symbols: (a) indirectly heated cathode; (b) directly heated cathode.

the device. When a vacuum tube is forward-biased (plate positive with respect to the cathode), electrons move from cathode to plate. This is equivalent to saying that conventional current flow is from plate to cathode in a vacuum diode. When the diode is reverse-biased, there is no current, because the plate cannot emit electrons and the cathode cannot emit holes. (Refer to the section on symbols and conventions in Chapter 0 for an explanation of conventional current and electron flow.)

The *I-V* characteristic of a typical vacuum diode is shown in Figure 2.7. The current reaches a limiting value, called *saturation,* when all the emitted electrons are being drawn to the plate. Except for the saturation region, which is not shown in the figure, we can use the ideal diode in series with a battery and a resistance as being the equivalent. The only thing that makes this equivalent circuit different from the one we use for the *PN* junction diode is the fact that the battery voltage might be zero or it might be connected in the opposite direction. In the literature dealing more thoroughly with vacuum diodes, you may see a family of curves, that is, several curves drawn on the same set of axes. This is done to show, for example, how the *I-V* characteristic varies with filament temperature. It is not important, at this point, to do a detailed study of vacuum diodes.

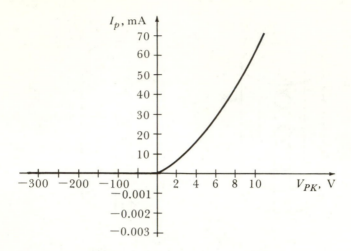

Figure 2.7 Vacuum-diode *I-V* characteristics.

2.7 GAS DIODES

In general, we find that for large-current applications the ordinary vacuum diode has considerably more forward voltage drop than does a suitable *PN* junction diode for the same application. In order to lower the forward voltage drop and also to increase the current carrying capacity, some tubes have a controlled amount of gas, rather than a high vacuum, contained within the envelope. Mercury vapor is commonly used in this application. Another type of gas-filled tube, the cold-cathode voltage regulator, will be discussed later.

2.8 DIODE SPECIFICATIONS AND RATINGS

An ideal diode would be suitable for any circuit application requiring diode action. Since ideal diodes are nonexistent, we must have some way to specify the diode characteristics needed in a particular application.

In many diode applications we find that there are two diode ratings of greatest importance. These ratings are the *forward-current* rating and the *reverse-voltage* rating. Before discussing these ratings, and others as well, let us take a preliminary look at Figures 2.8 and 2.9. These are manufacturers' data sheets on *PN* junction diodes of vastly different ratings and potential applications. Figure 2.8 is for a switching diode. A probable application for such a diode would be in

Silicon
Diodes

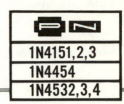

1N4151,2,3		
1N4454		
1N4532,3,4		

This family of General Electric silicon signal diodes are very high speed switching diodes for computer circuits and general purpose applications. These diodes incorporate an oxide passivated planar structure. This structure makes possible a diode having high conductance, fast recovery time, low leakage, and low capacitance combined with improved uniformity and reliability. These diodes are contained in two different packages;
double heat sink miniature package and milli-heat sink package and are electrically the same as their equivalent types in each of the different packages. (see page two for groupings of electrically equivalent types in each of the packages).

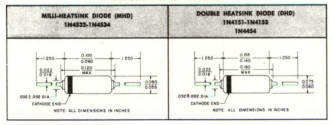

PLANAR EPITAXIAL PASSIVATED
with Controlled Conductance

MILLI-HEATSINK DIODE (MHD) 1N4532-1N4534	DOUBLE HEATSINK DIODE (DHD) 1N4151-1N4153 1N4454
NOTE: ALL DIMENSIONS IN INCHES	NOTE: ALL DIMENSIONS IN INCHES
Dissipation: 500mW @ 25°C free air **Derate:** 2.85mW/°C for temp. above 25°C amb. based on max. T₁ = 200°C	**Dissipation:** 500mW @ 25°C free air **Derate:** 2.85mW/°C for temp. above 25°C amb. based on max. T₁ = 200°C

FEATURES	1N4151 1N4454 MHD618 1N4532	1N4152 1N4153 1N4533 1N4534
Reverse Recovery Time of 4 nanoseconds maximum	●	●
Min.-Max. V_F specified at 6 Forward Current Levels		●
Capacitance of 2 pF maximum	●	●
Power Dissipation to 500 mW	●	●
Power Dissipation to 250 mW		
Meets all MIL-S-19500 requirements	●	●

Figure 1

HEATSINK SPACING FROM END OF DIODE BODY	STEADY STATE THERMAL RESISTANCE °C/mW*		POWER DISSIPATION AT 25°C mW†	
	MHD	DHD	MHD	DHD
.062"	.230	.250	760	700
.250"	.319	.319	550	550
.500"	.380	.380	460	460

*See Figure 5 for thermal resistance for short pulses.
†This power rating is based on a maximum junction temperature of 200°C.

Figure 2.8 Specifications for a family of high-speed switching diodes. (Courtesy of General Electric Company.)

absolute maximum ratings: (25°C) (unless otherwise specified)

	1N4454 1N4532	1N4151 MHD618	1N4152 1N4533	1N4153 1N4534	DHD MHD
Voltage					
Reverse	50	50	30	50	Volts
	MHD & DHD Units				
Current					
Average Rectified	150				mA
Recurrent Peak Forward	450				mA
Forward Steady State DC	200				mA
Peak Forward Surge (1 μsec. pulse)	2000				mA
Power					
Dissipation	500				
Temperature					
Operating	←——— −65 to +200 ———→				°C
Storage	←——— −65 to +200 ———→				°C

---

electrical characteristics: (25°C) (unless otherwise specified)

		1N4454* 1N4532		1N4151 MHD618		1N4152 1N4533		1N4153 1N4534		
		Min.	Max.	Min.	Max.	Min.	Max.	Min.	Max.	
Breakdown Voltage										
($I_R = 5\mu A$)	B_V	75		75		40		75		Volts
Forward Voltage										
($I_F = 100\mu A$)	V_F					0.490	0.550	0.490	0.550	Volts
($I_F = 250\mu A$)	V_F					0.530	0.590	0.530	0.590	Volts
($I_F = 1mA$)	V_F					0.590	0.670	0.590	0.670	Volts
($I_F = 2mA$)	V_F					0.620	0.700	0.620	0.700	Volts
($I_F = 10mA$)	V_F		1.00			0.700	0.810	0.700	0.810	Volts
($I_F = 20mA$)	V_F					0.740	0.880	0.740	0.880	Volts
($I_F = 50mA$)	V_F				1.00					Volts
Reverse Current										
($V_R = 30V$)	I_R						50			nA
($V_R = 30V, T_A = +150°C$)	I_R						50			μA
($V_R = 50V$)	I_R		100		50				50	nA
($V_R = 50V, T_A = +150°C$)	I_R		100		50				50	μA
Reverse Recovery Time										
($I_F = I_R = 10mA, I_{rr} = 1mA,$ Figs. 9 & 10)	t_{rr}		4		4		4		4	nsec.
($I_F = 10mA, V_R = 6V, I_{rr} = 1mA,$ $R_L = 100$ ohms, Figs. 9 & 10)	t_{rr}		2		2		2		2	nsec.
Peak Forward Voltage†	V_{peak}		3.0							Volts
Capacitance										
($V_R = 0V$)‡	C_o		2		2		2		2	pF
Stored Charge (Note 1)										
($I_F = 10mA$)§	Q_s		32		32		32		32	pC

*MIL type available
†50mV peak square wave, 0.1 usec. pulse width, 5 to 100 kHz repetitive rate, generator $t_r = 30$ nsec.
‡Capacitance as measured on Boonton Model 75A capacitance bridge at a signal level of 50 mV and a frequency of 1 MHz at $V_R = 0$ volts.
§Stored Charge as measured on B-Line Electronics Model QS-3 stored charge meter. Pulse amplitude = 5 volts, pulse width = 50 nsec., rise time = 0.4 nsec., source impedance = 10 ohms.

Figure 2.8 (continued)

the signal circuits of a digital computer. The family of diodes represented by the data given in Figure 2.9 are *high-current* diodes. These diodes might be specified for a circuit that must convert large amounts of electric energy from an alternating current (ac) type to a direct current (dc) type.

One of the important diode ratings is the forward-current rating. This rating tells us the maximum current the diode can safely

Westinghouse

**High Power
Silicon Rectifier Diodes
1N4587- 1N4596** ①

Up to 150 Amperes Half-Wave Average
Repetitive Peak Reverse Voltages to
1400 Volts

Application

Westinghouse JEDEC Series 1N4587 rectifiers are hermetically sealed high-voltage silicon rectifier diodes for all types of power applications. They may be used with proper cooling to provide average currents up to 150 amperes per rectifier diode, with maximum peak reverse voltage ratings up to 1400 volts.

These devices have been specifically designed for high-current, high-voltage industrial and military applications. The glazed ceramic body provides a long surface creepage path.

The many features of these rectifier diodes include high ambient temperature operation (up to 200°C junction temperature), long life, high efficiency, good regulation, rugged construction, small size, and freedom from thermal fatigue.

Exclusive Westinghouse CBE (compression Bonded Encapsulation) provides a THERMAL FATIGUE-FREE device by eliminating solder joints.

Features

- Low Thermal Impedance
- High Surge Current Capability
- CBE Construction

Dimensions in inches (And Millimeters)

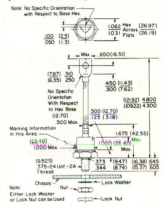

Weight: Approximately 3.5 ounces (99 gms)

DO-8 Case

Maximum Ratings and Characteristics

Blocking State (TJ = 200°C)

	Symbol	1N4587	1N4588	1N4539	1N4590	1N4591	1N4592	1N4593	1N4594	1N4595	1N4596
		JEDEC Number ①									
* Repetitive Peak Reverse Voltage, Volts	VRRM	100	200	300	400	500	600	800	1000	1200	1400
* Non-repetitive Peak Reverse Voltage, Volts	VRSM	200	300	400	525	650	800	1050	1300	1600	1800
* Max. Allowable d-c Blocking Voltage, Volts	VR	100	200	300	400	500	600	800	1000	1200	1400
* Max. (fca) ②, Reverse Current at Rated VRRM 150 Amperes Avg. Forward Current, Single Phase @ Tc = 110°C, ma	IR(AV)	9.5	9.5	9.0	9.0	8.0	6.5	5.5	4.5	4.0	3.5

Conducting State (TJ = 200°C)	Symbol	All Types
* Max. (fca) ②, Forward Current @ Tc = 110°C, Amperes	IF (AV)	150
RMS Forward Current, Amps	IF (RMS)	236
* Max. Peak ½ Cycle Surge Current (at 60 Hz) (Under Load) Amps	IFSM	3,000
I²t for Fusing (at 60 Hz Half-Wave), Amps²—Sec	I²t	37,200
* Max. Forward Voltage Drop @ 150 Amperes Average, Tc = 110°C, Peak Volts	VFM	1.35
Thermal Characteristics		
* Oper. Junction Temp. Range, °C	TJ	−65 to +200
* Storage Temperature Range, °C	Tstg	−65 to +200
Max. Thermal Resistance, °C/Watt Junction to Case	ReJC	.35
Case to Sink, Lubricated Mounting Surface	Recs	.12

Mechanical Characteristics

Finish: Nickel-plated case to maintain low contact resistance and to prevent corrosion.

Mounting: Recommended stud mounting torque (Clean threads); 12-15 ft. lb.

* JEDEC registered parameters.

① Order reverse polarity units by designating 1N4587R, etc.

② Full cycle average (measured with a d-c meter).

Figure 2.9 Specifications for high-current-capacity diode family. (Courtesy of Westinghouse Electric Corporation.)

High Power
Silicon Rectifier Diodes
1N4587– 1N4056 ①

Up to 150 Amperes Half-Wave Average
Repetitive Peak Reverse Voltages to
1400 Volts

Electrical Characteristics

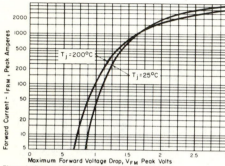

Figure 1. Forward current vs. Forward voltage.

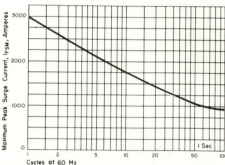

Figure 2. Maximum allowable surge current at rated load conditions.

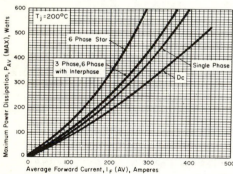

Figure 3. Power dissipation vs. Average forward current.

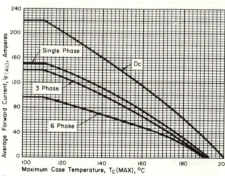

Figure 4. Forward current vs. Case temperature.

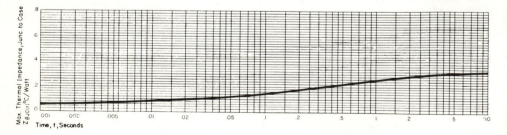

Figure 2.9 (continued)

conduct when forward-biased, that is, in the *on* state. In many instances the diode will alternate between the *on* and *off* states depending on the circuit waveforms. In anticipation of this usage, we find that the forward-current rating is given as the average current. Note that the data sheets also give surge-current capability.

The current-handling rating of a diode is largely a function of its heat-dissipating ability. It is the average current through the diode that determines the average power that the diode must dissipate as heat. The circumstances controlling heat dissipation are all-important in determining the current rating of a diode. The same diode may have vastly different current ratings when, for instance, printed-circuit-board mounting and heat-sink mounting are compared. The physical size of the diode package contributes greatly to its heat-dissipating ability and thus to its current rating. The temperature of the surrounding air will affect the current rating of a diode. When comparing diode ratings, make sure that you note the ambient or case temperature specified by the manufacturer.

There are certain things that can be done to improve the heat-dissipating ability of a diode, or any other component. One of the most common methods is to mount the component so that it is in intimate contact with a heat sink. A *heat sink* is any device that absorbs heat from a hot object and transfers that heat to another object or location. Many commonly used heat sinks transfer the heat from a hot diode (or other component) into the surrounding atmosphere. A few of the many available hot-component-to-atmosphere-type heat sinks are shown in Figure 2.10. The heat sinks shown are able to transfer heat to the nearby atmosphere because of their color (black), the high thermal conductivity of the material (aluminum), and the unusual shapes, which provide exceptionally large amounts of surface area.

The efficiency of the heat sinks shown in Figure 2.10 is dependent on several factors. One obvious factor is the temperature of the surrounding atmosphere. Another factor is the physical configuration of the heat sink. If the mounting of the heat sink is such that the cross-sectional view of Figure 2.10 is a top view, natural convection will cause an upward airflow past the fins which will improve the heat-dissipating ability of the sink as compared with any other mounting configuration. A third factor has to do with airflow. In addition to the natural convection mentioned previously, forced circulation of air can be used to advantage. If forced air is used, the physical configuration must be such as to take advantage of the direction of airflow. Figure 2.11 is a graphical comparison of the same heat sink under the conditions of natural convection and forced air movement.

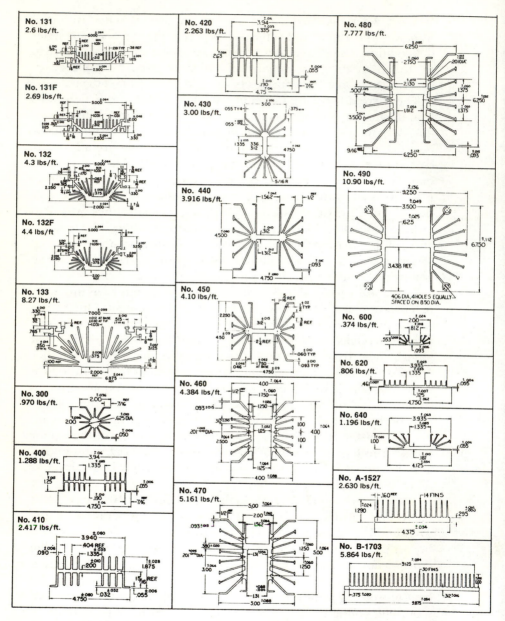

Figure 2.10 Heat sinks; a few of the many types and configurations available. (Courtesy of Wakefield Engineering, Inc.)

FOR POWER SEMICONDUCTORS

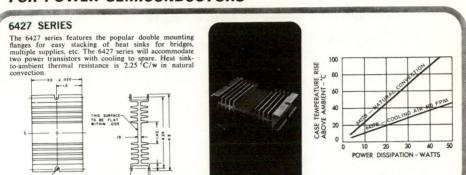

Figure 2.11   A graphical representation of heat-sink temperature rise comparing natural convection versus forced air for a particular heat sink. (Courtesy of Thermalloy, Inc.)

At times it will be necessary or desirable to operate a diode (or other component) at a temperature higher than that used to write the specifications of the device. In such a case, we must derate the device because of its limited heat-dissipating ability. Derating curves are commonly used to find ratings at the higher temperatures. A typical derating curve is part of the manufacturer's specifications shown in Figure 2.9. A quick check will show that the 300-A-rated diode must be derated to 125 A dc at 160°C; and the rating continues to decrease until zero is reached at a temperature of 175°C. Sometimes the derating curve is a straight line. In these cases, the manufacturer may say something like "for temperatures above 100°C derate linearly to 175°" or perhaps "for temperatures greater than 75°C derate at 3 A (or 20 mA or 5 mW, and so forth) per degree Celsius."

Another diode rating of great importance is the reverse voltage rating. Many times this is referred to as the peak inverse voltage or PIV rating. The PIV rating tells us how much voltage the diode can withstand in the reverse direction before the diode fails to have diode characteristics. Another name for this quantity is reverse breakdown voltage. When the PIV of a diode is exceeded, the diode will conduct in the reverse direction. The breakdown may or may not be destructive, depending on the diode and other circuit considerations.

There are many other possible diode ratings. These ratings may be all-important in certain applications and of no importance in other applications. Among these ratings are the following:

1. Surge current
2. Power dissipation
3. Operating temperature range
4. Storage temperature range
5. Junction capacitance
6. Reverse current
7. Forward voltage
8. Reverse recovery time

These and other ratings are given in many forms and under various test conditions. Thus we will not attempt to speak of each one individually. The data sheets of Figures 2.8 and 2.9 should be an indication of typical practice by manufacturers.

Diodes are available in a broad range of current and voltage ratings. The ratings put various restrictions on the size and shape of the diode packages. From Figure 2.12 we can get some appreciation of the various package sizes and configurations available. Most, but not all, of the packages shown contain single or multiple diodes. In some cases a package may contain two diodes with only three leads brought out. A full-wave rectifier—to be considered later—could use such a three-terminal device.

2.9   DIODES IN CIRCUIT APPLICATIONS

The forward-conducting and reverse-nonconducting characteristic of diodes is very useful in many circuits. Waveshaping, rectification, clamping, switching, and digital computer logic functions are some of the functions performed with circuits that commonly contain diodes. A few waveshaping examples will suffice here inasmuch as rectification will be covered thoroughly along with power supplies in Chapter 3. Special purpose diodes and other applications will be covered in Chapter 4.

EXAMPLE 2.1   Find the current waveform and the output voltage waveform for the circuit shown in Figure 2.13(a).

SOLUTION   The most straightforward and logical approach is to work the problem at several different instants of time. Since there are no time constants involved in the circuit, we can mentally stop

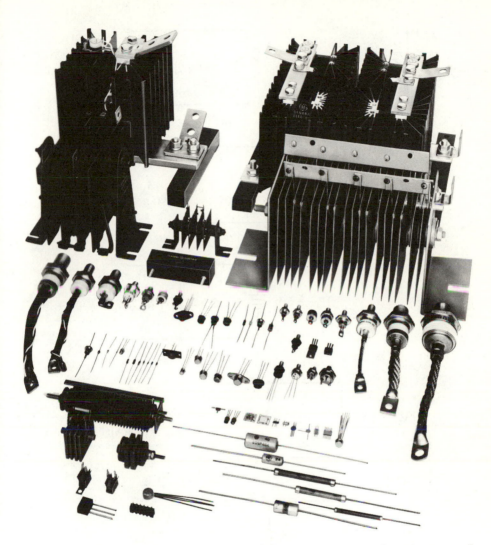

Figure 2.12 Diodes of various ratings and packaging styles. (Courtesy of General Electric Company.)

the action at any instant of time. We can replace the signal voltage source with a battery whose voltage magnitude and polarity are the same as the instantaneous signal voltage at the time in question. We can replace the diode symbol with its equivalent circuit. With the action stopped, we can solve for any dc voltage or current of interest. If we solve for enough of these instantaneous dc values, at appropriate instants of time, we can construct accurately the waveform of any voltage or current throughout the circuit. The intermediate steps in

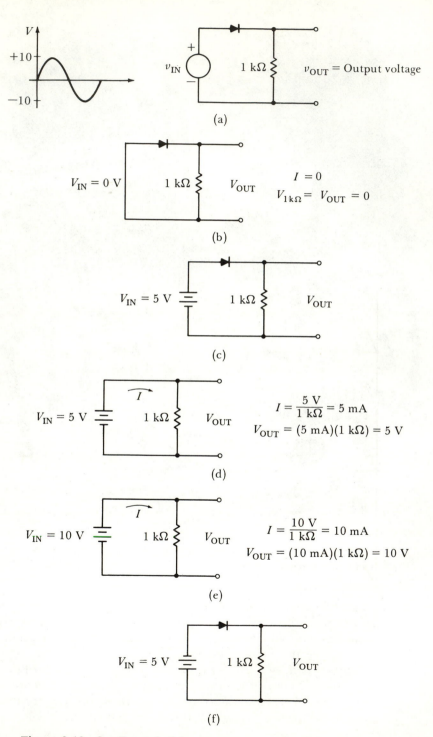

Figure 2.13   See Example 2.1.

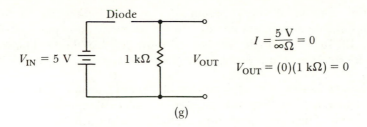

$$I = \frac{5 \text{ V}}{\infty \Omega} = 0$$

$$V_{OUT} = (0)(1 \text{ k}\Omega) = 0$$

(g)

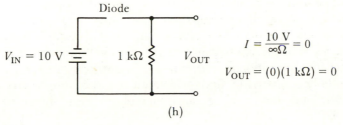

$$I = \frac{10 \text{ V}}{\infty \Omega} = 0$$

$$V_{OUT} = (0)(1 \text{ k}\Omega) = 0$$

(h)

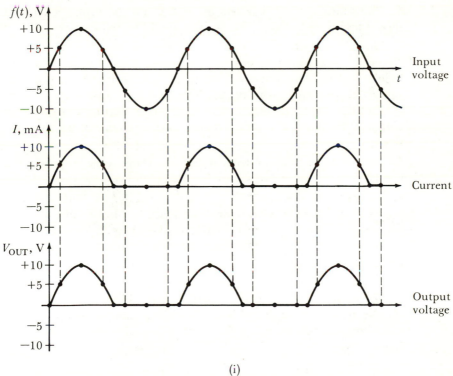

(i)

Figure 2.13 (continued)

the solution of the problem, that is, the equivalent circuit and appropriate calculations, are shown as the various parts of Figure 2.13. In this example, we are assuming an ideal diode. The steps in the solution of the problem are as follows.

1. Let us solve the problem for all instants of time when the input voltage is zero. The voltage source has been replaced with a short circuit. The circuit and the solution are shown in Figure 2.13(b).

2. The solution at all instants of time when the input voltage is 5 V will be based on the circuit shown in Figure 2.13(c). The instantaneous 5 V is replaced by a 5-V battery. The battery is of such a polarity as to attempt to cause current flow in the clockwise direction. Since the conventional current flow is in the direction of the arrow of the diode symbol, the diode will conduct and be a short circuit. The circuit, and solved voltage and current values, is shown in Figure 2.13(d).

3. The solution at all instants of time when the input is 10 V is shown in Figure 2.13(e). The results should be self-explanatory.

4. Find the solution at all instants of time when the input is −5 V. The equivalent circuit is shown in Figure 2.13(f). The battery is of such a polarity as to attempt to cause conventional current flow against the arrow of the diode. The diode is an open circuit to current flow in this direction. Thus the circuit becomes as shown in Figure 2.13(g). The solved voltage and current values are also shown.

5. Now let us find the solution at all instants of time when the input is −10 V. The circuit and solved values are shown in Figure 2.13(h).

6. By collecting the calculated data graphically, we obtain several points on the desired waveforms. It should be obvious at this point what the complete waveforms will look like when all data points are connected. The applicable constructed waveforms are shown in Figure 2.13(i).

We should now have a reasonably good idea of how to solve simple circuits containing diodes. Our next example will be slightly more complicated because a nonideal equivalent will be used for the diode.

EXAMPLE 2.2   Find the voltage waveform across the diode for the circuit shown in Figure 2.14(a). Use an equivalent circuit for the diode. The diode characteristic is also shown.

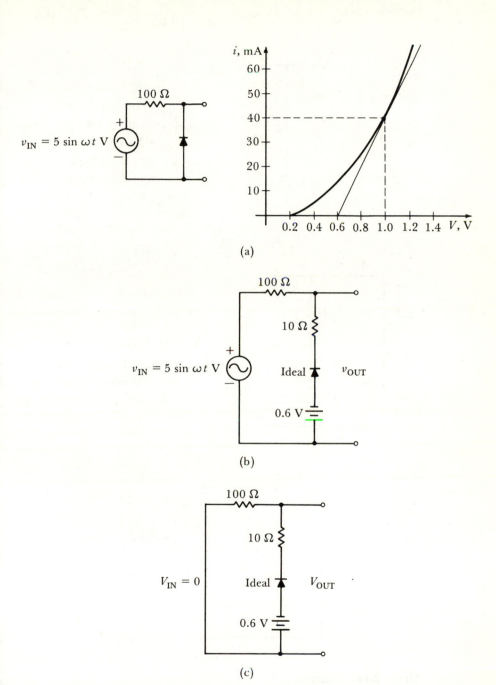

(a)

(b)

(c)

Figure 2.14   See Example 2.2.

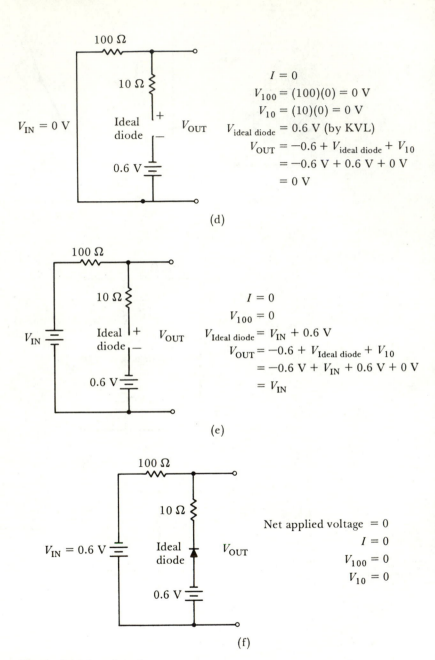

$$I = 0$$
$$V_{100} = (100)(0) = 0 \text{ V}$$
$$V_{10} = (10)(0) = 0 \text{ V}$$
$$V_{\text{ideal diode}} = 0.6 \text{ V (by KVL)}$$
$$V_{\text{OUT}} = -0.6 + V_{\text{ideal diode}} + V_{10}$$
$$= -0.6 \text{ V} + 0.6 \text{ V} + 0 \text{ V}$$
$$= 0 \text{ V}$$

(d)

$$I = 0$$
$$V_{100} = 0$$
$$V_{\text{Ideal diode}} = V_{\text{IN}} + 0.6 \text{ V}$$
$$V_{\text{OUT}} = -0.6 + V_{\text{Ideal diode}} + V_{10}$$
$$= -0.6 \text{ V} + V_{\text{IN}} + 0.6 \text{ V} + 0 \text{ V}$$
$$= V_{\text{IN}}$$

(e)

Net applied voltage $= 0$
$$I = 0$$
$$V_{100} = 0$$
$$V_{10} = 0$$

(f)

Figure 2.14 (continued)

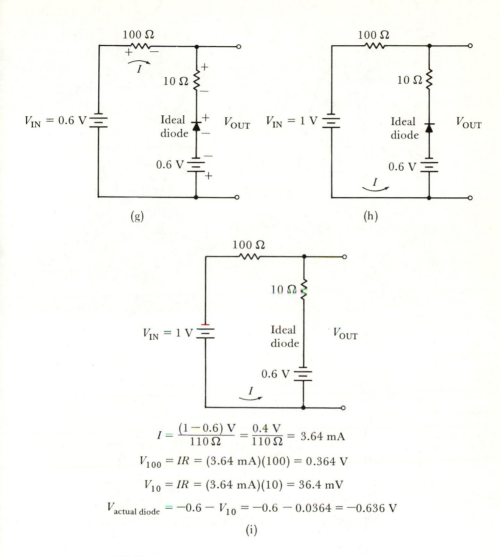

$$I = \frac{(1-0.6)\ V}{110\ \Omega} = \frac{0.4\ V}{110\ \Omega} = 3.64\ mA$$

$$V_{100} = IR = (3.64\ mA)(100) = 0.364\ V$$

$$V_{10} = IR = (3.64\ mA)(10) = 36.4\ mV$$

$$V_{\text{actual diode}} = -0.6 - V_{10} = -0.6 - 0.0364 = -0.636\ V$$

(i)

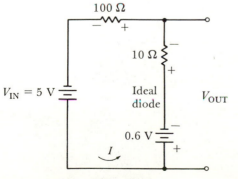

$$I = \frac{4.4\ V}{110\ \Omega} = 40\ mA$$

$$V_{100} = IR = (40\ mA)(100) = 4\ V$$

$$V_{10} = IR = (40\ mA)(10) = 0.4\ V$$

$$V_{\text{actual diode}} = -0.6 - V_{10} = -1\ V$$

(j)

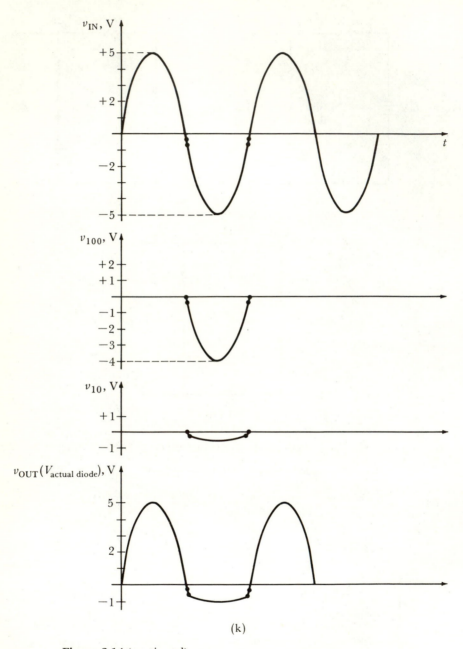

(k)

Figure 2.14 (continued)

SOLUTION

1. First we need to find the component values for the diode equivalent circuit. By inspection of the construction on the diode characteristic in Figure 2.14(a), we can see that the turn-on voltage is 0.6 V. The dynamic resistance is

$$\frac{V}{I} = \frac{0.4 \text{ V}}{40 \text{ mA}} = 0.01 \text{ k}\Omega = 10 \text{ }\Omega$$

Substituting the equivalent circuit for the diode, we obtain the circuit shown in Figure 2.14(b), which we will analyze to determine the waveshapes.

2. At all instants of time when the input is zero volts, the circuit is as shown in Figure 2.14(c). There is a net voltage that will attempt to cause conventional current flow in the clockwise direction. Thus the circuit becomes as shown in Figure 2.14(d), where we also have the solution for voltage and current quantities.

3. At all instants of time when the input is zero or positive, the results of step 1 will repeat because the diode is back-biased. This more general solution is considered in this step. The circuit and the solution is shown in Figure 2.14(e). V_{IN} is to be interpreted as any positive voltage.

4. At all instants when the input voltage is −0.6 V, the circuit and analysis of Figure 2.14(f) results. By Kirchhoff's voltage law we can find the voltage across the ideal diode. First let us arbitrarily define a direction for current flow even though we know that the magnitude of current flow is zero. Then we can put polarity signs on all voltages around the loop. For a clockwise current, the circuit diagram with polarities is as shown in Figure 2.14(g). Starting at the bottom left of the circuit diagram and going clockwise, we get the following equation from Kirchhoff's voltage law:

$$-0.6 \text{ V} - V_{100} - V_{10} - V_{\text{ideal diode}} + 0.6 \text{ V} = 0$$

Substituting for known quantities, we have

$$-0.6 \text{ V} - 0 - 0 - V_{\text{ideal diode}} + 0.6 \text{ V} = 0$$

$$-V_{\text{ideal diode}} = 0$$

$$V_{\text{ideal diode}} = 0$$

We are now in a position to calculate the voltage across the actual diode.

$$V_{\text{OUT}} = V_{\text{actual diode}} = -0.6 \text{ V} + V_{\text{ideal diode}} + V_{10}$$

$$= -0.6 + 0 + 0$$

$$= -0.6 \text{ V}$$

An easier method in this case would have been to observe that the output voltage is the sum of the input voltage and V_{100}. Since we know both, we conclude that

$$V_{OUT} = -0.6 + 0$$
$$V_{OUT} = -0.6V$$

5. At all instants of time when the input is -1 V, the circuit of Figure 2.14(h) applies. A net voltage of 0.4 V exists that attempts to cause current flow in the counterclockwise direction. The ideal diode is a short circuit to current flow in this direction, so the circuit can be simplified as shown in Figure 2.14(i).

6. Based on the situation found to exist in step 5, we should be able to draw some general conclusions about the circuit that will help us solve for voltages and currents for other input voltages more negative than 0.6 V.

(a) As long as the input voltage is more than 0.6 V negative, the circuit in Figure 2.14(i) will always apply.

(b) The net applied voltage will always be the input voltage minus the 0.6 V in such a direction as to cause current in the counterclockwise direction.

(c) The magnitude of current will be the net applied voltage divided by the total resistance, which in this case is 110 Ω.

$$I = \frac{\text{net voltage}}{110 \ \Omega}$$

(d) $V_{100} = IR = I(100)V$.

(e) $V_{10} = IR = I(10)V$.

(f) $V_{\text{actual diode}} = -0.6 - V_{10} = -0.6 - I(10)$.

7. The solution at instants of time when the input is peak negative $(-5$ V), an example of the general case discussed in step 6, should give us enough information to determine the total waveform. The circuit is shown in Figure 2.14(j) with calculated data.

8. By collecting all data points we are now able to construct the desired waveforms. The construction is shown in Figure 2.14(k).

2.10   COMMENTS ON EXAMPLES

The problems we have just worked certainly do not exhaust all the possible configurations. Based on the principles used in the preceding solutions, we will hopefully be able to solve problems of this type in any configuration. Some circuits, for example, may have batteries or voltage sources somewhere around the loop. Additional batteries are handled in exactly the same manner as the diode equivalent circuit battery and the instantaneous input voltage. The component placement order around the loop may be different from that in the examples and in addition the waveform of interest may be across any single component or any combination of components. As long as we have no capacitors or inductors in the circuit to cause the circuit to have a time constant, and therefore voltages and currents dependent on time, we can mentally cause the input to stop at some various instants of time and do a dc analysis at the various chosen instants of time.

We should now be in a position to solve most non time-constant circuits containing a diode. With experience, you will find you need to solve the circuit for fewer and fewer time instants in order to arrive at the complete waveform of interest.

EXERCISES

QUESTIONS

Q2.1 Why will a *PN* junction conduct in one direction and not in the other?

Q2.2 The best equivalent circuit we use to describe a *PN* junction diode has a dynamic resistance, a battery, and an ideal diode. What circumstances might make it practical to use a less-complex equivalent circuit?

Q2.3 What similarities and what differences do you note when comparing a *PN* junction diode with a set of switch contacts?

Q2.4 What are the advantages of *PN* junction diodes as compared with vacuum tube diodes?

Q2.5 What consequences may result from exceeding the forward-current rating of a *PN* junction diode?

Q2.6 What consequences may result from exceeding the peak-inverse-voltage (PIV) rating of a *PN* junction diode?

Q2.7 How and why will a heat-sink mounting extend the forward-current rating of a *PN* junction diode?

Q2.8 What are the factors that influence the efficiency of heat sinks?

Q2.9 Why is it a valid technique for us to solve the examples of this chapter by stopping the action at various time instants?

PROBLEMS

P2.1 Draw the piecewise-linear *I-V* characteristic of the *PN* junction diodes that are specified as follows:

(a) $V_f = 0.3$ V, $r = 10$ Ω (b) $V_f = 0.6$ V, $r = 0$

P2.2 Draw the piecewise-linear *I-V* characteristic of the *PN* junction diodes that are specified as follows:

(a) $V_f = 0.7$ V, $r = 50$ Ω (b) $V_f = 0$, $r = 20$ Ω

P2.3 What are the absolute minimum ratings for average forward current (I_0) and reverse voltage (PIV) necessary for the diode in the circuit of Figure 2.15?

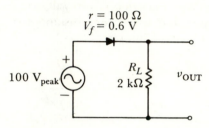

Figure 2.15   See Problems 2.3 and 2.7.

P2.4 Find numerical values for r and V_f for a diode whose characteristics are shown in Figure 2.16 if it is known that the on current will be near 60 mA.

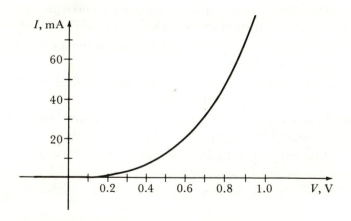

Figure 2.16   See Problems 2.4 and 2.5

P2.5 Find numerical values for r and V_f for a diode whose characteristics are shown in Figure 2.16 if the on current will be between 10 and 20 mA.

P2.6 For the circuit shown in Figure 2.17,

    (a) Draw the current waveform.

    (b) Draw the v_{OUT} waveform.

    (c) Draw the diode voltage waveform.

    (d) Draw the diode current waveform.

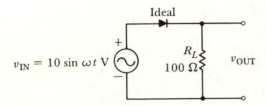

Figure 2.17 See Problem 2.6.

P2.7 For the circuit shown in Figure 2.15,

    (a) Draw the current waveform.

    (b) Draw the v_{OUT} waveform.

    (c) Comment on errors to be expected if an ideal diode were assumed.

P2.8 For the circuit of Figure 2.18,

    (a) Draw the current waveform. (b) Draw the v_{OUT} waveform.

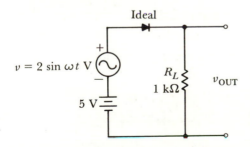

Figure 2.18 See Problems 2.8 and 2.9.

P2.9 Repeat Problem 2.8 with the diode reversed.

P2.10 Draw the i and v_{OUT} waveforms for the circuit of Figure 2.19 for the case of an ideal diode.

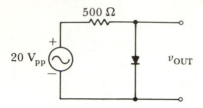

Figure 2.19 See Problems 2.10 and 2.11.

P2.11 Draw the i and v_{OUT} waveforms for the circuit of Figure 2.19 if $r = 200 \ \Omega$ and $V_f = 1$ V.

P2.12 Referring to the circuit of Figure 2.20,

   (a) Draw the i waveform.   (b) Draw the v_{OUT} waveform.

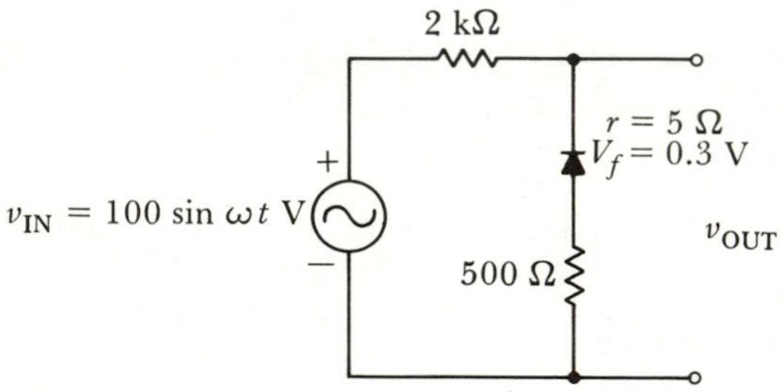

Figure 2.20 See Problem 2.12.

P2.13 For the circuit of Figure 2.21,

   (a) Draw the current waveform.

   (b) Draw the voltage-output waveform.

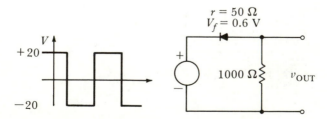

Figure 2.21 See Problems 2.13 and 2.14.

P2.14 Repeat Problem 2.13 with the diode reversed.

P2.15 Draw the following waveforms for the circuit of Figure 2.22:

   (a) Circuit current   (b) v_{AB}

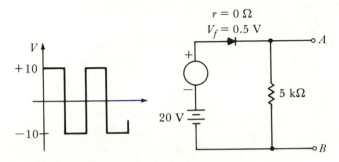

Figure 2.22   See Problems 2.15 and 2.16.

P2.16 Repeat Problem 2.15 with reversed diode.

P2.17 Referring to the circuit of Figure 2.23,

   (a) Draw the I waveform.   (b) Draw the v_{FG} waveform.

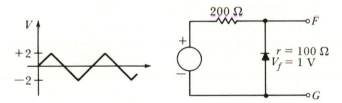

Figure 2.23   See Problems 2.17 and 2.18.

P2.18 Repeat Problem 2.17 with reversed diode.

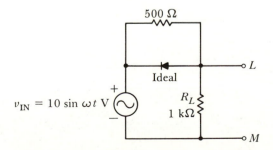

Figure 2.24   See Problems 2.19 and 2.20.

P2.19 A circuit containing a diode is shown in Figure 2.24.

    (a) Draw the i_{diode} waveform.

    (b) Draw the i_{500} waveform.

    (c) Draw the v_{LM} waveform.

P2.20 Repeat Problem 2.19 for the case where the diode is reversed in the circuit.

P2.21 Referring to the circuit diagram of Figure 2.25,

    (a) Draw the i_{1000} waveform. (b) Draw the v_{OUT} waveform.

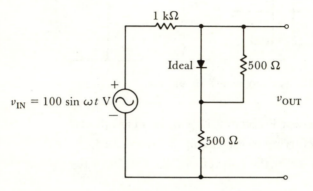

Figure 2.25   See Problems 2.21 and 2.22.

P2.22 Repeat Problem 2.21 for the diode is reversed.

CHAPTER 3
RECTIFIERS AND POWER SUPPLIES

In many industrial and electronic applications, a direct-current source of electric energy is needed. Most electric power easily made available to us by the power companies is of the alternating-current type. The circuit that performs the ac-to-dc conversion is called a *rectifier circuit.*

The output of a rectifier circuit is a pulsating dc voltage (or current) waveshape. However, a pulsating direct current is not good enough for many electronic applications. A circuit that will take a pulsating dc waveshape and convert it into a smooth dc waveshape (as smooth as desired) is called a *filter circuit.* When speaking of a power-supply circuit for electronics, we generally are referring to a circuit that performs both a rectification and a filtering function.

A power-supply circuit is an in-between circuit and a signal-conditioning circuit. It is positioned between an ac voltage source and a circuit that requires a smooth dc voltage or current. Its purpose is to convert and condition an ac signal into a smooth dc signal. The user of the dc signal energy from a power-supply circuit is called a *load* device (or component, or circuit). As examples of load circuits, we might include amplifier circuits, a radio receiver, or a dc motor. For simplicity, a dc load circuit is often shown as an equivalent resistance.

The name of power supply for the circuits we are studying is somewhat inappropriate, since the generator at the power company plant supplies the electric power that results from an energy conversion. Nevertheless, we will abide by common usage and call the circuits discussed in this chapter either rectifier circuits or power-supply circuits.

3.1 HALF-WAVE RECTIFIER CIRCUITS

The circuit shown in Figure 3.1 is a half-wave rectifier. The generator is considered to be an ac voltage source. The diode may be considered ideal or an equivalent may be substituted depending on the application and the required accuracy. The resistance R_L is used to represent the device to which we are delivering electric power. R_L may be representing a heating element, the resistance of the coil on a dc relay, a dc motor, a radio receiver, or any one of a multitude of other devices.

The waveforms of Figure 3.1(b) and (c) show the current and the output voltage associated with a half-wave rectifier. The waveforms may be easily obtained in the same manner as that used in the

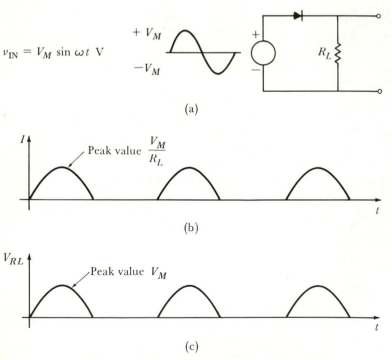

(a)

(b)

(c)

Figure 3.1   Half-wave rectifier: (a) circuit; (b) current waveform; (c) voltage waveform across R_L.

examples of Chapter 2. When an ideal diode is assumed, an even easier method consists of two parts as follows:

1. When the input voltage is positive, the diode is a short circuit. The peak current is the peak applied voltage V_M divided by the total resistance R_L. The peak output voltage is the peak current times the resistance R_L, which simplifies to V_M. The peak output voltage must be V_M, since the load and voltage source are in parallel when the diode is a short circuit.

2. When the input voltage is negative, the diode is an open circuit. The circuit current is zero and the load voltage is also zero.

If the diode in the circuit of Figure 3.1 were reversed, the diode would be a short circuit for negative inputs and an open circuit for positive inputs. The circuit is called a half-wave rectifier because only one of the two input waveform alternations appear at the output of the circuit.

The diode used in this circuit must have a sufficient current rating and a PIV rating of at least the peak of the input voltage waveform V_M.

3.2 FULL-WAVE RECTIFIER CIRCUITS

The circuit shown in Figure 3.2 is a commonly used circuit for obtaining full-wave rectification. The transformer is arranged so that the voltage at point A is opposite in polarity from the voltage at point B with respect to the center tap. The polarity relations are shown in Figure 3.2(b) and (c). The voltages v_{AN} and v_{BN} are of equal magnitude.

If you have not yet studied transformers, you will need to accept as fact that we can, and do, quite easily obtain the necessary voltage magnitudes and the necessary polarity relations for full-wave rectification. Since transformers are covered in other courses, we will not cover them here.

The full-wave rectifier circuit shown in Figure 3.2 is in reality two half-wave rectifier circuits driving the same load circuit with opposite polarity. The relationship between the two input voltages causes one half-wave circuit to operate during one of the input alternations and the other half-wave circuit to operate during the second input alternation. There is a pulse of current and of voltage in the output waveform for each and every input alternation.

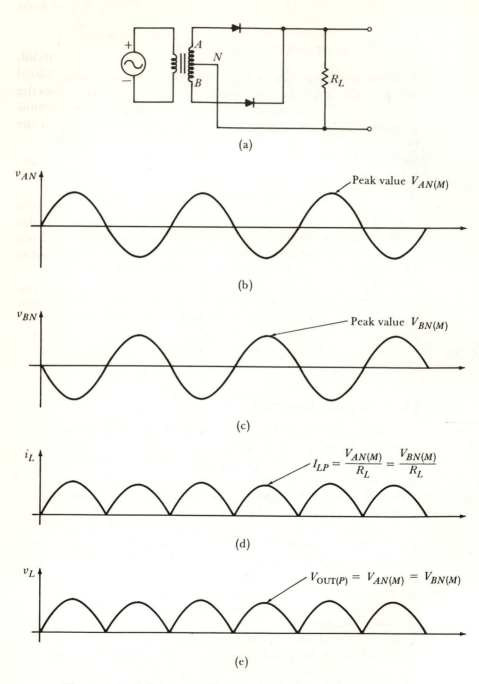

Figure 3.2 Full-wave rectifier: (a) circuit; (b) voltage waveform at point A wrt point N; (c) voltage waveform at point B wrt point N; (d) current waveform through R_L; (e) voltage waveform across R_L.

The reverse breakdown voltage rating of the diodes used in the full-wave rectifier circuit of Figure 3.2 must be at least $2V_M$. This is the case because each time the output voltage is at a peak value, one of the diodes is experiencing a peak of the opposite polarity at its opposite end. At this instant the diode in question has $+V_M$ at one end and $-V_M$ at the other end. Each voltage is expressed with respect to transformer center tap or circuit ground.

A transformerless full-wave rectifier circuit is possible. The bridge rectifier is a transformerless full-wave rectifier circuit that is quite widely used. Figure 3.3 shows two common schematic arrangements for a bridge rectifier circuit. Figure 3.4 shows the direction and path

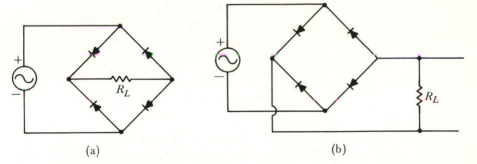

(a)   (b)

Figure 3.3 Bridge rectifier: (a) schematic representation; (b) alternate location of load R_L.

of current flow for each of the two input alternations. Since we note that the direction of current through the load resistance is the same during each alternation, we know that the load voltage is of the same polarity for each alternation of the input waveform.

The full-wave rectified waveforms shown in Figures 3.2 and 3.5 can be obtained on a point-by-point basis as we have previously done. (See examples in Chapter 2.) The full-wave rectified waveforms we have shown are for the ideal diode. If we know the characteristics of the actual diode to be used, we can substitute the equivalent circuit for the diode in the schematic diagram and solve for the current and voltage waveforms on a point-by-point basis. The waveforms for the actual diode will not normally be greatly different from that for the ideal diode.

The PIV rating of the individual bridge rectifier diodes must be at least the peak value of the input voltage.

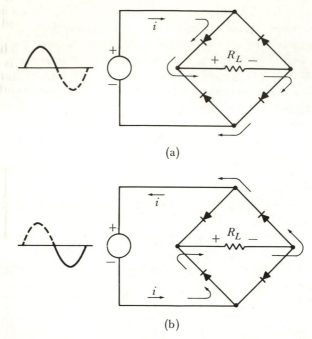

(a)

(b)

Figure 3.4   Bridge rectifier: (a) current direction during positive input alternation; (b) current direction during negative input alternation.

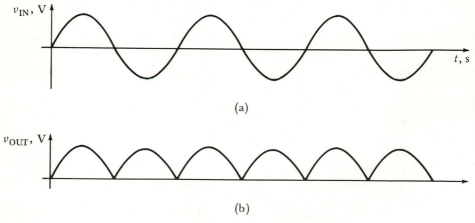

(a)

(b)

Figure 3.5   Full-wave bridge rectifier waveforms: (a) input-voltage waveform; (b) output-voltage waveform.

3.3 COMMENTS ON RECTIFIER CIRCUITS

The half-wave and full-wave rectifier circuits convert alternating current to direct current. The output direct current is not steady but pulsating. The current through, and the voltage across, the load resistance must be considered direct current since it is unidirectional and does not change polarity. Even though the polarity of the output is constant, the magnitude is not constant. The changing magnitude of the output is called ripple. An examination of Figures 3.1 and 3.2 will show that the ripple repetition frequency is different for the half-wave and full-wave cases. For a half-wave rectifier, the ripple frequency is the same as the input sine-wave frequency. For the full-wave rectifier, the output ripple frequency is twice the input sine-wave frequency. In many cases the higher ripple frequency is advantageous in that it can be removed more easily than can low-frequency ripple.

Each rectifier type has certain advantages and disadvantages. The half-wave rectifier and the bridge rectifier have an advantage over the transformer-type full-wave rectifier in that no transformer is required. The full-wave rectifiers have the advantage of double ripple frequency. The bridge rectifier has the advantage of being transformerless, but it has a disadvantage in that the input circuit and the output circuit cannot have a common connection. If, for instance, one side of the input to a bridge rectifier is grounded, neither side of the load resistance can be grounded.

3.4 RECTIFIER CONVERSION EFFICIENCY

We define conversion efficiency for rectifier circuits as the dc power delivered to the load divided by the total power input to the rectifier circuit. Conversion efficiency is usually represented by the Greek letter η (eta). We will shortly discover that half-wave rectifier circuits and full-wave rectifier circuits are not equal in conversion efficiency.

At this point it will be helpful to introduce a new analysis method to help us calculate conversion efficiency. It will also be useful in the study of power-supply filters. The Fourier series is a mathematical way of expressing nonsinusoidal waveforms in an infinite series consisting of one dc term and an infinite number of ac terms. Since Fourier analysis is normally studied at later points in the curriculum, it will be sufficient here to write several terms of the Fourier series for the half-wave and full-wave rectified waveforms. In general, the

Fourier series for a nonsinusoidal waveform may be written as

$$f(t) = \frac{a_0}{2} + a_1 \cos \omega_0 t + a_2 \cos 2\omega_0 t + a_3 \cos 3\omega_0 t + \cdots$$

$$+ b_1 \sin \omega_0 t + b_2 \sin 2\omega_0 t + b_3 \sin 3\omega_0 t + \cdots$$

$$= \frac{a_0}{2} + \sum_{n=1}^{\infty} (a_n \cos n\omega_0 t + b_n \sin n\omega_0 t)$$

The values $a_0, a_1, a_2 \ldots$ and $b_1, b_2, b_3 \ldots$ are constants whose values depend on the particular waveform being represented by the series. a_0 is an average quantity, whereas the remaining a_n and b_n quantities represent the peak values of sine or cosine alternating components. The constants a_n and b_n get smaller and finally approach zero as n gets large. There may be individual a_n or b_n values larger than the preceding value, but the trend must be toward smaller magnitudes for larger n.

In order to satisfy our needs, we will give the Fourier series for only the half-wave and full-wave rectified waveforms. In Figure 3.6 we have a full-wave rectified waveform and its associated Fourier series. Figure 3.7 shows two half-wave rectified waveshapes and the associated Fourier series representation of each. The waveshapes of Figure 3.7 are identical except for the placement of the time axis. Note that the Fourier series is different when the time reference is changed.

The Fourier series has been mentioned so that the reader will realize that the unidirectional output of a rectifier can, and should, be considered as the sum of dc and ac components. Since a rectifier is intended to convert ac power to dc power, the only desired

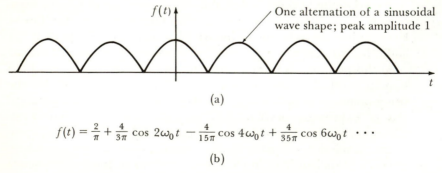

$$f(t) = \frac{2}{\pi} + \frac{4}{3\pi} \cos 2\omega_0 t - \frac{4}{15\pi} \cos 4\omega_0 t + \frac{4}{35\pi} \cos 6\omega_0 t \cdots$$

(b)

Figure 3.6   Full-wave rectifier: (a) full-wave rectified waveform; (b) Fourier series representation.

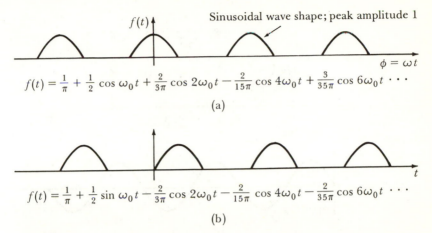

Figure 3.7 Half-wave rectifier: (a) waveshape and Fourier series for rectified cosine wave; (b) waveshape and Fourier series for rectified sine wave (time-shifted cosine wave).

component at the output of a rectifier circuit is the dc component that is the first term in the Fourier series.

A review of the commonly used current and voltage units of measure will be of help to us in calculating rectifier conversion efficiency. As you probably remember, the root mean square (rms) voltage and current units were introduced such that time-varying signals with, for instance, 100 V rms would have the same average heating effect (power) as would 100-V direct current across the same load resistance. Although you may have used rms quantities only with sinusoidal waveforms up to the present time, it should be stated—and remembered—that rms quantities include the effects of both ac and dc quantities. When we use rms quantities to calculate power, we are calculating total power contributed by both ac and dc components.

In terms of the Fourier series, a power calculation using rms quantities includes the power contributed by each and every term of the Fourier series representation of the particular waveform involved. If we use the average values associated with a particular waveform to calculate power, we are in fact considering only the power contributed by the dc term, the first term, in the Fourier series representation of the waveform.

We are now in a position to draw a diagram of a rectifier circuit and to calculate the conversion efficiency. In Figure 3.8 we have a typical rectifier circuit showing current and voltage waveforms at the input and output. The diode is represented here as having a series

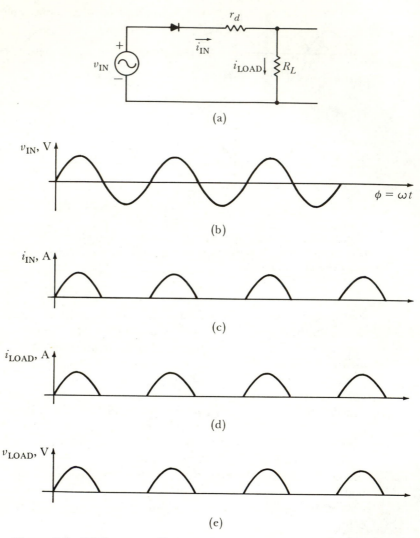

Figure 3.8   Half-wave rectifier: (a) circuit diagram; (b) input-voltage wave-form; (c) input-current waveform; (d) load-current waveform; (e) load-voltage waveform.

resistance r_d but no turn-on voltage. The defining equation for power conversion efficiency (η) can be written as follows:

$$\eta = \text{conversion efficiency} = \frac{\text{dc power out}}{\text{dc + ac power in}}$$

$$= \frac{\text{dc power out}}{\text{total power in}}$$

Since it is obvious for the circuit shown in Figure 3.8(a) that only the current waveform is the same for input and output circuits, and since the efficiency defining equation involves both input and output circuits, we find it convenient to write the efficiency equation in terms of the rectified current waveform. In this context, we use the symbol I to indicate the rectified current waveform and follow the symbol with a subscript to indicate if we are referring to the rms or to the average value of the current. The efficiency equation thus becomes

$$\eta = \frac{(I_{av})^2 \, R_L}{(I_{rms})^2 \, (R_L + r_d)}$$

In order to complete the numerical computation for the circuit of Figure 3.8, we will need to find I_{av} and I_{rms} for half-wave rectified signals. The average current over one cycle is the area under the curve for one complete cycle divided by the base of that complete cycle. It is probably easiest to think of one complete cycle as being 2π rad. Thus we will draw the waveform and write the defining equations in terms of the angle ϕ. This has been done in Figure 3.9. The area under the curve is to be found by integration.

$$A = \text{area under curve} = \int_0^{2\pi} f(\phi) \, d\phi$$

$$= \int_0^{\pi} I_M \sin \phi \, d\phi + \int_{\pi}^{2\pi} (0) \, d\phi$$

$$= I_M \int_0^{\pi} \sin \phi \, d\phi = I_M(-) \cos \phi \Big|_0^{\pi}$$

$$= -I_M(\cos \pi - \cos 0) = -I_M(-1 - 1) = I_M(-2)$$

$$= +2I_M$$

$$I_{av} = \frac{\text{area}}{\text{base}} = \frac{2I_M}{2\pi} = \frac{I_M}{\pi}$$

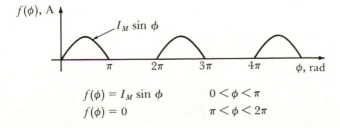

$$f(\phi) = I_M \sin \phi \qquad 0 < \phi < \pi$$
$$f(\phi) = 0 \qquad \pi < \phi < 2\pi$$

Figure 3.9 Half-wave rectified waveshape and mathematical description in terms of angle ϕ.

Note that the value of average current just calculated is exactly the same as the first term (dc term or average term) in the Fourier series for a half-wave rectified current wave-form of peak value I_M.

The results of the preceding integration, that is, the fact that the area under one arch of a sine or cosine wave is twice the peak value, should be noted carefully and remembered. It will save you time and work in the future.

We also need to find I_{rms} for the half-wave rectified current wave. Root mean square calls for squaring the waveform, finding the mean (average) height of the squared waveform, and then taking the square root of the averaged squared waveform. Figure 3.10 shows the original and the squared waveforms in terms of the angle ϕ. We can find the average height of the squared waveform by finding the area under the curve of the squared waveform and then dividing by the base. The area under the curve is found by integration as follows:

$$A = \text{area} = \int_0^{2\pi} f(\phi)\, d\phi = \int_0^{\pi} I_M^2 \sin^2 \phi \, d\phi + \int_\pi^{2\pi} (0)\, d\phi$$

$$= I_M^2 \int_0^{\pi} \sin^2 \phi \, d\phi = I_M^2 \int_0^{\pi} (\tfrac{1}{2} - \tfrac{1}{2} \cos 2\phi) d\phi$$

$$= I_M^2 \int_0^{\pi} \tfrac{1}{2} \, d\phi - I_M^2 \int_0^{\pi} (\tfrac{1}{4}) \cos 2\phi(2) \, d\phi$$

$$= I_M^2 (\tfrac{1}{2}) \, \phi \Big|_0^{\pi} - I_M^2 (\tfrac{1}{4}) \sin 2\phi \Big|_0^{\pi}$$

$$= I_M^2 (\tfrac{1}{2}) \, (\pi - 0) - I_M^2 (\tfrac{1}{4}) (\sin 2\pi - \sin 0)$$

$$= \frac{I_M^2 \pi}{2} - 0$$

$$= \frac{I_M^2 \pi}{2}$$

The average height of the squared waveform is the area divided by the base, which is

$$\text{Average } I^2 = \frac{\text{area}}{\text{base}} = \frac{I_M^2 \pi}{2} \frac{1}{2\pi} = \frac{I_M^2}{4}$$

The root mean square is the root of the averaged squared waveform, which is

$$I = \sqrt{\frac{I_M^2}{4}} = \frac{I_M}{2}$$

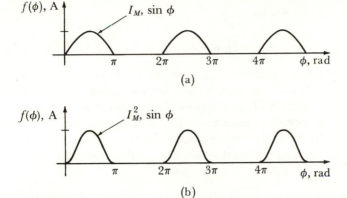

Figure 3.10 Rectified waveforms: (a) half-wave rectified current waveform; (b) square of rectified current waveform.

Efficiency η is then calculated in terms of the rectified current wave as follows:

$$\eta = \frac{\text{dc power out}}{\text{total power in}}$$

$$= \frac{I_{\text{dc}}^2 R_L}{I_{\text{rms}}^2 (R_L + r_d)}$$

$$= \frac{I_{\text{av}}^2 R_L}{I_{\text{rms}}^2 (R_L + r_d)} = \frac{\left(\dfrac{I_M}{\pi}\right)^2 R_L}{\left(\dfrac{I_M}{2}\right)^2 (R_L + r_d)}$$

$$= \frac{4}{\pi^2} \frac{R_L}{R_L + r_d}$$

If the values of R_L and r_d had been given, we could easily calculate the numerical value of the conversion efficiency as a decimal or as a percentage. The efficiency is probably most often given as a percentage.

To improve efficiency, it would be of great interest to know what values of R_L and r_d result in the highest possible conversion efficiency. It can be seen by inspection of the preceding equation for efficiency that efficiency is a maximum, and is independent of R_L, when r_d approaches zero resistance. Efficiency can be increased by making r_d small as compared with R_L. Efficiency is an absolute maximum when r_d is zero. In that case we have

$$\eta_{\max} \text{ (for half-wave)} = \frac{4}{\pi^2}$$

$$\eta_{\max} = 0.405 = 40.5\%$$

An approach similar to the one just concluded for the half-wave rectifier will now be undertaken for the full-wave rectifier circuit shown in Figure 3.11. During one alternation, the current is through D_1, its associated resistance r_d, R_L, and the top half of the transformer. During the other input-voltage alternation, current is through D_2, its associated resistance r_d, R_L, and the bottom half of

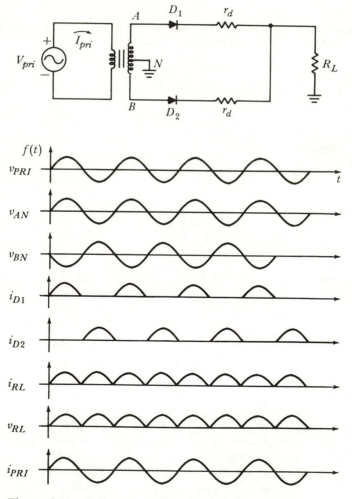

Figure 3.11   Full-wave rectifier circuit and waveforms.

the transformer secondary. The result is current through the load during each input alternation, as shown in the waveforms of Figure 3.11. Note that input current ($I_{pri} = I_{D1} + I_{D2}$) and output (load) current are of the same waveshape, which enables us to calculate both input and output power based on the same full-wave rectified current waveform. The diodes are assumed identical and therefore the series resistances r_d are equal and have the same symbol. In our calculations we will not be able to distinguish between the two r_d values.

To find the total input power, we will need to know the rms value of the full-wave rectified current waveform. This of course includes both ac and dc power components. During one alternation, input power is delivered to R_L and the top r_d value; and during the other alternation, input power is supplied to R_L and the bottom r_d value. Thus, we can say that the input power is always delivered to the sum of R_L plus r_d.

Let us now calculate the rms and average values of current in terms of the peak value of the full-wave rectified current waveform. Figure 3.12 shows the full-wave and squared full-wave waveforms in terms of ϕ. The average current is the area under one arch divided by the base of one arch. Remembering that the area under one arch of a sine wave is twice the peak value, we obtain the following relation for average current:

$$I_{av} = \frac{\text{area}}{\text{base}} = \frac{2I_M}{\pi}$$

The rms current is, of course, the square root of the average of the squared waveform over one repetitive waveshape. We will need to

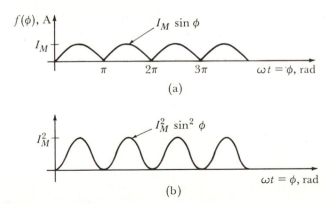

Figure 3.12   Rectified waveforms: (a) full-wave rectified current waveform; (b) square of rectified current waveform.

use calculus to find the area under one arch of the squared waveform to obtain the average value of the squared waveform.

$$A = \text{area} = \int_0^\pi f(\phi)\, d\phi = \int_0^\pi I_M^2 \sin^2 \phi\, d\phi$$

$$= \frac{I_M^2 \pi}{2}$$

$$I_{av}^2 = \frac{\text{area}}{\text{base}} = \frac{I_M^2 \pi}{2}\frac{1}{\pi} = \frac{I_M^2}{2}$$

The square root of the averaged squared waveform is, of course, the rms current we are seeking.

$$I_{rms} = \sqrt{\frac{I_M^2}{2}} = \frac{I_M}{\sqrt{2}}$$

The conversion efficiency of a full-wave rectifier can be written using the results of the preceding mathematical manipulations.

$$\eta \text{ (for full-wave)} = \frac{\text{dc power out}}{\text{total power in}} = \frac{(I_{av})^2 R_L}{I_{rms}^2 (R_L + r_d)}$$

$$= \frac{(2I_M/\pi)^2 R_L}{(I_M/\sqrt{2})^2 (R_L + r_d)} = \frac{(4)(2)}{\pi^2}\frac{R_L}{R_L + r_d}$$

$$= \frac{8R_L}{\pi^2 (R_L + r_d)}$$

If we know R_L and r_d, in a specific problem, we can compute the numerical value of conversion efficiency.

As in the case of the half-wave rectifier, maximum efficiency occurs when r_d is minimum. The absolute maximum efficiency that can be obtained with a full-wave rectifier occurs when r_d is zero. Then we have

$$\eta_{max} \text{ (for full-wave)} = \frac{8}{\pi^2}$$

$$\eta = 0.81 = 81\%$$

3.5   POWER-SUPPLY FILTER CIRCUITS

In many cases the amplitude of the nonconstant output voltage of a rectifier circuit is inappropriate. The ripple cannot be tolerated. An example of an application where high ripple voltages cannot be tolerated is in the dc supply to vacuum tubes or transistors in a radio receiver. Any ripple will be audible as hum. We are thus faced with the task of removing the ripple voltage from the output of a rectifier circuit so that the load receives very little ripple-voltage magnitude. The circuit used to minimize the ripple voltage is called a power-supply filter circuit. We shall study several types of circuits that are commonly used as power-supply filters.

3.6 SHUNT-CAPACITANCE FILTER

The addition of a capacitor in parallel with the load resistance as shown in Figure 3.13 is a simple, and common, filtering circuit. In this circuit, as well as in all others we use in our study of filtering, we assume that the diode is ideal.

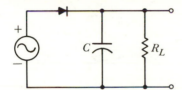

Figure 3.13   Shunt-capacitor filter circuit.

Consider the circuit of Figure 3.13 and the resulting waveform of Figure 3.14(c). During the first half of the positive-input alternation, the capacitor will charge to the peak input voltage. This must be so since the diode is a short circuit causing the voltage source, the capacitor, and the load resistor to be connected in parallel. Thus the load and input-voltage waveshapes must be identical during this time interval.

During the second half of the positive-input alternation, the situation is considerably different. As the input voltage decreases from a peak value toward zero, the capacitor will attempt to discharge. However the capacitor cannot discharge by the same path through which it charged since the diode will not allow current in the reverse direction. The capacitor must discharge through the load resistor. The capacitance value is normally chosen so that the discharge time constant is relatively long. Under these conditions we find that as the input voltage changes from peak positive value toward zero, it will decrease faster than the capacitor can discharge through R_L. Thus the diode will be reverse-biased. This will turn the diode off, and load current will be supplied by the filter capacitor. The capacitor

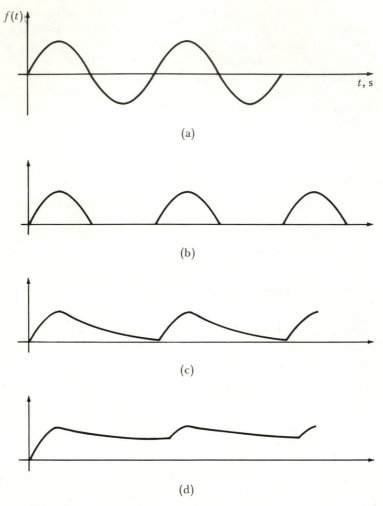

$f(t)$

t, s

(a)

(b)

(c)

(d)

Figure 3.14   Rectifier and filter waveforms: (a) input-voltage waveform; (b) output-voltage waveform without filter capacitor; (c) output-voltage waveform with medium R-C time constant; (d) output-voltage waveform with long R-C time constant.

can supply current by discharging through the load resistance. The load voltage during this discharge interval will be the familiar decaying exponential waveform.

In Figure 3.14 we see the waveforms associated with a half-wave rectifier circuit and note what effect the R-C time constant has on the ripple content of the output voltage. The waveform of Figure 3.14(b) can be considered as the output waveform when the R-C

time constant is very short. The waveform in Figure 3.14(d) is the well-filtered output when a long time constant is used in the filtering circuit. Theoretically we can make the ripple voltage increasingly small by using a long enough R-C time constant in the filter circuit.

Please note that the addition of a shunt filter capacitor puts additional requirements on the rectifier diode. The PIV rating must be at least $2V_M$. A look at the waveforms of Figure 3.14 should convince us of the validity of these statements. Comparing (a) and (d) we see that the output voltage is positive even at the instant the input is peak negative. It is at this instant that the diode is reverse-biased by $2V_M$.

Let us now consider the filtered output waveform carefully in order to calculate the ripple voltage. Although many authors attempt to calculate the rms ripple voltage, we will calculate the peak-to-peak ripple voltage because it seems more important. It can be approximated rather easily.

The output-voltage waveform of a shunt-capacitance filter circuit has two distinct parts. When the diode is shorted, the output is sinusoidal. When the diode is open, the output voltage is a decaying exponential. The waveform in Figure 3.15 is labeled with respect to the equations that describe the sinusoidal part and the exponential part of the wave. Note that the equation for the exponential is correct only if we consider time is zero at point a on the waveform. Time reference zero is at the origin for the sine wave and at point a for the exponential. If we know the voltage at point a and the voltage at point b, we have in fact the peak-to-peak ripple voltage as

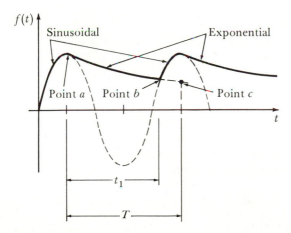

Figure 3.15 Output-voltage waveform for half-wave shunt-capacitance filter circuit.

the difference between these two voltages. The voltage at point a is, by inspection, V_M, that is, the peak value of the input waveform. We could find the voltage at point b by solving the sinusoidal equation and the exponential equations simultaneously, being careful to consider the time references correctly. Let us conserve our energy and time, however, by noting that the voltage at point c is not much different from the voltage at point b and can be calculated easily. We will therefore assume that the peak-to-peak ripple is the difference between the voltages at point a and point c, even though we know we are introducing some error into the calculations. Before condemning the making of error-inducing approximations, let us recall that the simultaneous solution of transcendental functions is not easy. In this case, it is a trial-and-error (iterative) solution. The voltage at point b is

$$v_{\text{OUT}} = V_M \epsilon^{-t_1/R_L C}$$

We have chosen to approximate this voltage with the voltage at point c, which is

$$v_{\text{OUT}} = V_M \epsilon^{-T/R_L C}$$

We make this particular approximation because the numerical value for T is readily available in most problems. The peak-to-peak ripple voltage is then

$$V_{\text{ripple(pp)}} = V_M - V_M \epsilon^{-T/R_L C}$$

The error introduced by this particular approximation is pessimistic in that the actual ripple will be somewhat less than the value you calculate.

EXAMPLE 3.1   A 100-V peak 60-Hz sine wave of voltage is applied to a half-wave rectifier circuit with a simple shunt-capacitor filter circuit. The load resistance is 500 Ω. Find the peak-to-peak ripple voltage if the capacitance is 50 μF.

SOLUTION   First draw the circuit and the waveforms anticipated both with and without the filter circuit. Then do the necessary mathematics to determine the maximum and minimum output voltage. The peak-to-peak ripple is the difference. The circuit and the applicable waveforms are shown in Figure 3.16.

By inspection of the output-voltage waveform, we find that

$$V_{\text{OUT}}(\text{max}) = 100 \text{ V}$$

We now need to solve the exponential part of the waveform at a time of $T = \frac{1}{60}$ to find the (approximate) value of the minimum voltage.

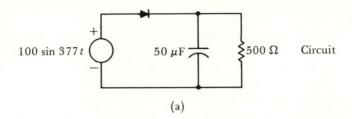

Circuit

(a)

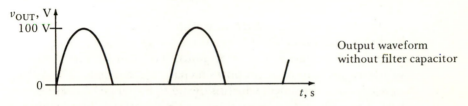

Output waveform
without filter capacitor

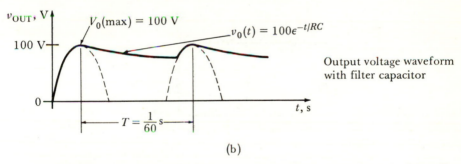

Output voltage waveform
with filter capacitor

(b)

Figure 3.16 Circuit and waveforms for Example 3.1. (a) Circuit; (b) applicable waveforms.

$$V_{OUT}(min) = 100\epsilon^{-T/RC}$$

$$= 100\epsilon^{-0.667}$$

$$= (100)(0.513) = 51.3 \text{ V}$$

Since we now know the maximum and minimum values of output voltage V_{OUT}, we can calculate the peak-to-peak ripple voltage as the difference.

$$V_{ripple(pp)} = V_{OUT}(max) - V_{OUT}(min)$$

$$100 - 51.3 = 48.7 \text{ V (pp)}$$

Note that this basic problem could be worded in several ways. For example, you could be given the permissible ripple and asked to find R, C, or input frequency with all other necessary facts given.

3.7   SERIES-INDUCTANCE FILTER

Another simple single-component filter is the series-inductance filter. Since an inductance offers an opposition (that is, reactance) to
the ac components of the rectified waveform but not to the dc
component, it is placed in series between the source and the load. An
inductor used in a filter application is called a *choke*. A series-
inductance (choke) filter circuit and waveforms are shown in Figure
3.17.

An explanation of the current waveform shown in Figure 3.17 can
be given in terms of the *I-V* relations for an inductor. The voltage
across an inductance is $v = L(di/dt)$. This can be interpreted in
words as "an inductance opposes a change in current." The voltage
v_L across an inductance, which is owing to a change of current, is of
such a polarity as to oppose the change in current. The magnitude

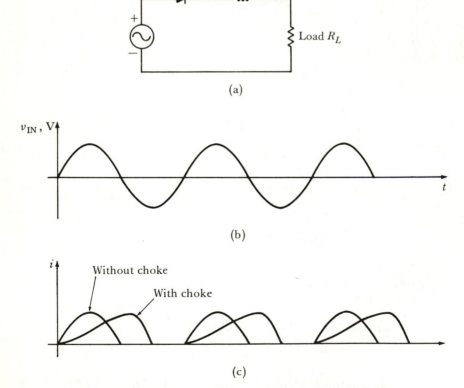

(a)

(b)

(c)

Figure 3.17   Half-wave rectifier with series-inductance filter circuit: (a)
circuit; (b) input-voltage waveform; (c) load-current waveform.

of the voltage v is determined by the rate at which the current is changing. Without the choke, we have a normal half-wave rectified waveform, as shown in Figure 3.17(c). With the choke, we find that the current buildup is slower because of the opposition to current change by the inductance. When the current would normally decrease to zero because of the input wave doing so, we find that the inductance opposes this decreasing change also and delays the fall to zero. Thus the current waveform shown can logically be expected.

The simple series choke filter gives some filtering action and lowers the ripple somewhat. In the process of filtering, we see that the current conduction angle is greater than 180°. It is possible to use this type of filter with a full-wave rectifier and obtain greater than 180° conduction through each diode and full 360° conduction in the load. The full-wave rectifier with series-inductance filter, and the associated waveforms, is shown in Figure 3.18.

Calculating the ripple voltage at the output of a series choke filter circuit is more difficult than calculating it for the shunt-capacitor filter. It is very difficult to find the mathematical expressions for the waveforms involved. The method of solution we will use makes use of the Fourier series. We will consider the sinusoidal voltage source in combination with the rectifying diode (or diodes) as being a

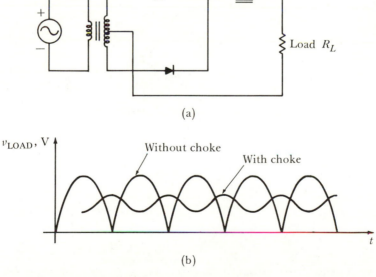

(a)

(b)

Figure 3.18 Full-wave rectifier with series-inductance filter circuit: (a) circuit; (b) load-current (and voltage) waveforms.

voltage source producing a rectified waveform of voltage. The rectified waveform is then applied across the series connection of inductance L and load resistance R_L. This sequence of events in the
solution is depicted in Figure 3.19.

The solution of the problem continues as follows. The rectified
voltage waveform is now represented by the Fourier series. We will
make use of the superposition theorem and solve for the output
voltage for each term of the Fourier series. The solution involves a
voltage divider consisting of L and R_L. We want to know what

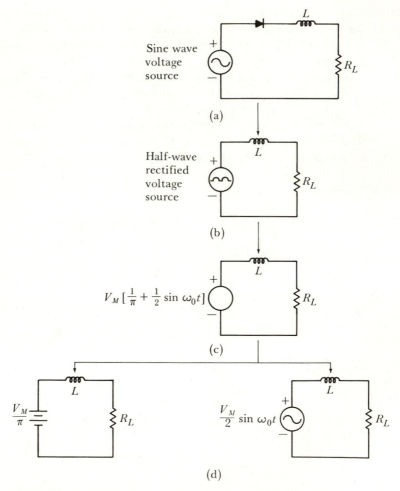

Figure 3.19 Method of solution for series-inductance filter circuit: (a)
original circuit; (b) assumed voltage-source equivalent; (c) Fourier representation of rectified voltage; (d) separate solution for each Fourier term.

portion of each Fourier voltage term appears across the load resistor R_L. This voltage-divider calculation will involve complex numbers, since the components involved are a resistance and a reactance. As we solve this type of problem, we find that the output voltage (across R_L) becomes very small for the higher-frequency terms of the Fourier series. Thus in most cases we obtain the approximate output by using only the first two terms of the Fourier series. The first term yields the dc output voltage and the second term yields the approximate ripple voltage.

Caution should be exercised with regard to the first ac term of the Fourier series that yields the ripple information. Since the Fourier coefficient represents the peak value of a sinusoidal voltage, the ripple voltage calculated will be in terms of peak units. In order to obtain the peak-to-peak voltage that we desire, we merely double the peak value.

It should be pointed out that this method of solution is not exact, even if we use each and every term of the Fourier series. The reason for the inexact solution is that we are using the Fourier series for a rectified sine wave when in fact the waveform across the series connection of choke and load is not exactly a rectified sine wave. The mathematical description of the exact waveform is difficult to obtain. We settle upon this method of solution as being a good approximation. It is relatively easy to use.

EXAMPLE 3.2   A full-wave rectifier has a 10-V peak across each half of the secondary of the transformer. The frequency is 60 Hz. The filter is a series inductance of 1 H and the load resistance is 1000 Ω. Estimate the peak-to-peak ripple voltage.

SOLUTION   First let us draw a diagram of the circuit involved. It is shown in Figure 3.20(a).

Now let us consider the appropriate Fourier series.

The Fourier series for a full-wave rectified waveform is

$$f(t) = v(t) = V_M \left(\frac{2}{\pi} + \frac{4}{3\pi} \cos 2\omega_0 t - \frac{4}{15\pi} \cos 4\omega_0 t + \cdots \right)$$

The dc output voltage can be obtained by inspection of the Fourier series. In fact, the entire dc term of the Fourier series appears at the output. This is so because (ideally) the series choke has no reactance to direct current. The dc part of Figure 3.19(d) helps to visualize this fact. The dc output voltage is

$$V_{OUT} = \frac{2V_M}{\pi} = \frac{(2)(10)}{\pi} = \frac{20}{\pi} = 6.35 \text{ V}$$

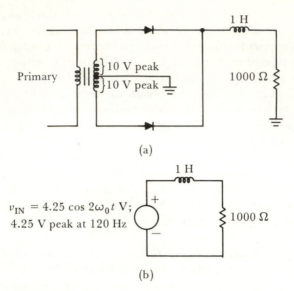

(a)

(b)

Figure 3.20   Circuit of Example 3.2.

From the Fourier series we see that the lowest-frequency ac component is

$$V_{out(m)} = \frac{4V_M}{3\pi} \cos 2\omega_0 t = 4.25 \cos 2\omega_0 t \text{ volts}$$

The frequency of the input voltage in this problem was given as 60 Hz. Thus f_0 is 60 Hz. Since we know that angular velocity

$\omega = 2\pi f$, we have

$\omega_0 = 2\pi f_0 = 2\pi(60) = 377$ rad/s

$2\omega_0 = 754$ rad/s

$2f_0 = 120$ Hz

For our purposes in calculating the lowest-frequency ac component at the load, we can consider the circuit to be as shown in Figure 3.20(b).

Now we will calculate X_L and the load current.

$$X_L = 2\pi f L = 6.28 \times 120 \times 1 = 753 \ \Omega$$

$$I_L \text{ (peak) at 120 Hz} = \frac{V}{Z} = \frac{4.25\underline{/0}}{1000 + j753} = \frac{4.25\underline{/0°}}{1250\underline{/37°}}$$

$$= 3.4\underline{/-37°} \text{ mA}$$

We will drop all phase angles inasmuch as they have no bearing on this problem.

I_L (peak) = 3.4 mA

The lowest-frequency ac voltage component across the load is then

V_L (peak)(ac) = 3.4 mA × $1k$ = 3.4 V

According to this method of calculation, the ripple voltage is $V_{L(pp)}$ = 6.8 V.

3.8 COMMENTS ON SIMPLE SERIES CHOKE FILTERS

It should be noted that to most people the method of solution we have used to solve for the ripple voltage of a series choke filter circuit is not very satisfying. The assumption we have made is not necessarily extremely close to fact. We have assumed, for example, that we have a rectified waveform (either full-wave or half-wave) driving the filter and load circuits. As a consequence of this assumption, we have used the Fourier series for a rectified waveform. In the no-filter case, we do in fact have the half-wave or full-wave rectified current and voltage waveforms. With the series choke filter, however, we do not have the assumed rectified waveforms (either current or voltage) at any point in the circuit. It is our assumption that the waveforms that actually exist will have a Fourier series not very different from the Fourier series for the rectified waveforms.

This theoretical inaccuracy would lead many to question the reliability of the results we obtained in the calculation of ripple voltages. In order to determine experimentally what percentage errors might be involved in this type of calculation, some circuits have been set up with the ripple voltage calculated and measured. A tabulation of calculated and measured ripple voltages for various circuit connections is shown in Table 3.1.

As can be seen from the entries in Table 3.1, the percentage error due to assumptions made for mathematical ease may result in relatively large errors in the results of our calculations. It is also interesting to note that the calculated values are all on the pessimistic side, that is, the calculated values are worse than the actual case.

All the comments about the mathematical methods and assumptions used in the calculation of ripple voltage for the simple choke filter circuit could easily be considered critical. Before we allow ourselves the luxury of criticizing the approach taken here, it would be wise to study the literature to become acquainted with other

Table **3.1**   Calculated and measured values of ripple voltage for simple series-inductance filter circuits

Circuit connection		Measured ripple	Calculated ripple	Error in calculated value
Half wave	400 Hz	5 V (pp)	6 V (pp)	20% too large
Half wave	60 Hz	160 V (pp)	168 V (pp)	5% too large
Full wave	60 Hz	5.1 V (pp)	6.8 V (pp)	33% too large

approaches. It is logical and valid to compare methods based on the following criteria:

1. The satisfaction associated with, and the understanding of, the theoretical background of the method.

2. The ease with which numerical results can be obtained.

3. The error introduced by the method.

After we have acquainted ourselves with the advantages and disadvantages of each method, we will be in a better position to choose the best method and to criticize the others. It is believed that most technology students will choose the methods presented here.

Although the simple series inductance (choke) filter can give fair filtering performance, especially with a full-wave rectifier, it is not commonly used for several reasons. The cost of a good high-inductance choke is considerable. The size and mass are comparatively great. The filtering performance is only fair; and the dc output can never exceed the average value of the unfiltered waveform in either the half- or full-wave rectified case. Any choke that can be built will have some dc resistance (which we have not considered) that will cause the dc output voltage to be somewhat lower than our calculations indicate.

The reason for studying the simple series inductance filter circuit is that chokes are commonly used in filter circuits in combination with capacitors. The ripple calculation techniques used here can also be applied to the LC filter.

3.9   THE *L*-SECTION FILTER

Sometimes inductors and capacitors are used together in a filter circuit, as in the configuration shown in Figure 3.21(a). This particular filter is called an *L-section filter.* These *L* sections may be cascaded into multisection filters, as shown in Figure 3.21(b).

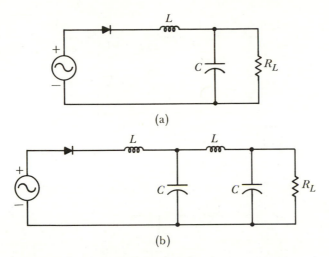

Figure 3.21 *L*-section filter. (a) Single *L*-section filter circuit; (b) two-stage *L*-section filter circuit.

Our chosen method of calculating the ripple voltage output for an *L*-section filter is quite similar to the method we used for the simple series inductance filter circuit. We will assume that the Fourier series for rectified waves is applicable to the waveform applied to the filter circuit.

In a properly designed *L*-section filter we find that the reactance of *C* is much less than the resistance of R_L, even at the fundamental Fourier frequency. Therefore in many cases we can neglect the value of R_L and assume that the voltage divider consists of *L* and *C*. The solution of the voltage-divider circuit will give us the ac voltage across capacitor *C*. Since R_L and *C* are in parallel, the ac voltage across *C* is the total ripple voltage at the load.

EXAMPLE 3.3   A half-wave rectifier circuit with an *L*-section filter has the following characteristics. The input is a 117-V rms 60-Hz sine wave. R_L is 500 Ω, *L* is 2 H, and *C* is 20 μF. Estimate the peak-to-peak ripple voltage across the load resistance.

SOLUTION   First we draw the circuit diagram as an aid to the solution of the problem. The circuit is shown in Figure 3.22(a). Our method of solution assumes that the circuit of Figure 3.22(b) applies. Since the lowest-frequency ac component predominates in the output ripple, we will consider only that frequency component in calculating ripple voltage. The lowest-frequency component is

$$v \,(60 \text{ Hz}) = (117)(1.414)(\tfrac{1}{2}) \cos (2\pi)(60)\,t$$

$$= 82.8 \cos (2\pi)(60)\,t$$

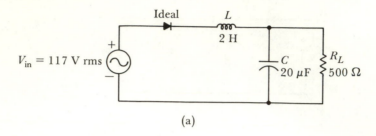

(a)

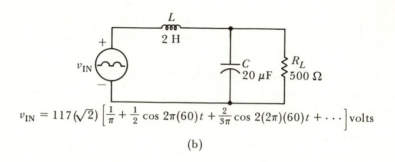

$$v_{IN} = 117 (\sqrt{2}) \left[\frac{1}{\pi} + \frac{1}{2} \cos 2\pi(60)t + \frac{2}{3\pi} \cos 2(2\pi)(60)t + \cdots \right] \text{volts}$$

(b)

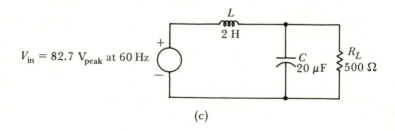

(c)

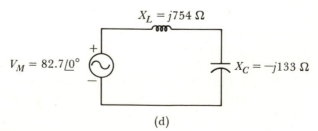

(d)

Figure 3.22   Circuit for Example 3.3.

As far as the 60-Hz component is concerned, we can simplify the circuit as shown in Figure 3.22(c).

Calculating the X_L and X_C values at 60 Hz, we obtain the following results:

$$X_L = 2\pi f L = 2\pi(60)(2) = 754 \ \Omega$$

$$X_C = \frac{1}{2\pi f C} = \frac{1}{(2\pi)(60)(20 \ \mu\text{F})} = 133 \ \Omega$$

Since we note that X_C is noticeably smaller than R_L, we assume that the parallel combination of X_C and R_L results in approximately $-j133 \ \Omega$. Our circuit is now simplified as in Figure 3.22(d).

We can now solve for the current and the output voltage.

$$I_{L_M} = \frac{V_M}{Z} = \frac{82.8\underline{/0°}}{j754 - j133} = \frac{82.8\underline{/0°}}{j621} \ \text{A}$$

Since we are not interested in phase information,

$$I_{L_M} = \frac{82.7}{621} = 0.133 \ \text{A}$$

$$V_{C_M} = I_M X_C = 0.133 \times 133 = 17.7 \ \text{V}$$

All voltage and current values were converted to peak quantities when we wrote the Fourier series. Thus, in terms of peak-to-peak quantities,

$$V_C \approx V_{RL} = V_0 = 2 \times 17.7$$

$$= 35.3 \ \text{V}_{\text{pp}}$$

Although not specifically asked for in this problem, the dc output voltage is

$$V_{\text{OUT}}(\text{dc}) = \frac{(117)(\sqrt{2})}{\pi}$$

$$= 52.7 \ \text{V}$$

The *L*-section filter is more common than the simple series inductor filter. It is more efficient at removing ripple voltage. The inductor, as before, may be large, heavy, and expensive.

Calculation of ripple voltage for the multiple-section filter is accomplished in a manner similar to that used for the single-section filter. If the X_L were considerably greater than X_C (by a factor of, say, 3 or more), then we would probably assume that the sections are completely independent, and no loading effects would be considered. If this assumption cannot be made, the calculation of ripple voltage by the common pencil-and-paper approach may be rather tedious. A digital computer is well-suited to such tedious work.

The L-section filter, as well as any other choke input filter, has a lower dc output level than does a comparable capacitance input filter. The first component following the diode is considered the filter input element.

3.10   π-SECTION FILTER

The filter circuit normally called a π-section filter is in reality an L-section filter with an added input capacitor. In Figure 3.23 we see a full-wave rectifier circuit with a π-section filter circuit. The π-section filter is commonly used. It can be very efficient in the removal of ripple voltage.

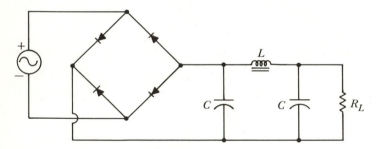

Figure 3.23   Full-wave bridge rectifier with a π-section filter.

Our method of calculating the ripple voltage present across the load for a circuit containing a π-section filter will require methods and assumptions we have previously used for the simple shunt-capacitance filter and the series inductance (or L-section) filter. Basically the method consists of determining the ripple voltage if only the input capacitor is present in the filtering circuit. Then we calculate the improvement to be obtained with the additional L-section components. It is presumed that the most beneficial way to explain the necessary procedures and assumptions is to work an example problem.

EXAMPLE 3.4   A full-wave bridge rectifier circuit is followed by a π-section filter. The input is 117-V rms at 60 Hz. The choke is 2 H and each capacitor is 20 μF. Estimate the peak-to-peak ripple voltage across a 1000-Ω load resistance.

SOLUTION First we must draw the circuit involved. This is shown in Figure 3.24(a).

The ripple calculation will be divided into two parts. The first part involves calculating the ripple if only C_1 is present. The circuit is then simplified as in Figure 3.24(b). The waveform across C and R under this circumstance will be as in Figure 3.24(c). From the waveform it can be seen that the peak value is 165 V. The minimum value (approximately) can be found by substituting $1/120$ s for t in the exponential equation.

$$V_{\text{OUT(min)}} \approx 165\epsilon^{-(1/120)/0.02} = 166\epsilon^{-0.417}$$

$$= (165)(0.66) = 109 \text{ V}$$

The peak-to-peak ripple voltage for the case of C_1 only in the circuit is then $165 - 109 \text{ V} = 56$ V.

The second part of the ripple calculation makes use of the results obtained for C_1 only in the circuit. We make the assumption that all the ripple voltage of the first part is at the lowest Fourier frequency, in this case 120 Hz. We can then draw an (approximately) equivalent circuit, as shown in Figure 3.24(d).

Let us now calculate values for X_L and X_C.

$$X_L = 2\pi fL = 2\pi \times 120 \times 2 = 1508 \ \Omega$$

$$X_C = \frac{1}{2\pi fC} = \frac{1}{2\pi \times 120 \times 20 \ \mu\text{F}} = 66 \ \Omega$$

Since X_C is much less than R_L, we will assume that X_C in parallel with R_L is $-j66 \ \Omega$. We can now calculate the ac component of current and the load voltage.

$$I_{\text{pp}} = \frac{V_{\text{pp}}}{Z} = \frac{56V_{\text{pp}}}{j1510 - j66} = \frac{56V_{\text{pp}}}{1442\underline{/90°}}$$

$$= 39.1 \text{ mA}_{\text{pp}}$$

Note that we dropped all phase information inasmuch as it is of no interest to us.

$$V_C \approx V_{RL} = V_{\text{OUT}} = IX_C = 39.1 \text{ mA} \times 66 \ \Omega$$

$$= 2.59 \text{ V}_{\text{pp}}$$

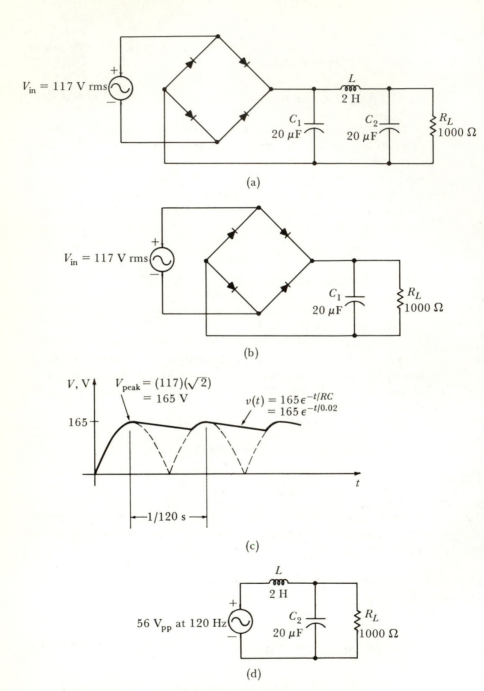

Figure 3.24 Circuit for Example 3.4.

3.11   COMMENTS ABOUT FILTER CIRCUITS

It is becoming more and more common to see filter circuits containing only shunt capacitance. This is because many transistor circuits require low voltages and because low-voltage high-capacitance capacitors are available and reasonably priced. The L-section and π-section filters are common but have the mutual disadvantage of having a large, heavy, and expensive choke.

Many filter circuits other than those specifically named and studied here can be built. In general, a filter circuit is to keep any ac components of the rectified wave from reaching the load. Any combination of inductors and capacitors will perform a filtering function if the inductors are placed in series with the load and the capacitors are placed in shunt. Any capacitance input circuit is expected to produce a higher dc output voltage level than is a similar choke input filter. Remember that for a capacitance input filter, the dc output can approach the peak value of the input; but for a choke input filter, the dc level can approach only the average value of the rectified input waveform.

3.12   RECTIFYING VOLTAGE-MULTIPLIER CIRCUITS

Diodes and capacitors can be connected in such a way as to obtain a dc output voltage that is a multiple of the peak value of the ac input voltage. Such circuits are called *rectifying voltage-multiplying* circuits or simply *voltage multipliers*. When the multiple is 2, 3, or 4, it is common to speak of voltage-doubler, voltage-tripler, or voltage-quadrupler circuits.

The circuit shown in Figure 3.25 is a half-wave voltage-doubler circuit. It is easy to analyze this circuit if we break it into two parts and analyze the parts separately.

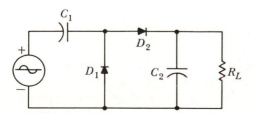

Figure 3.25   Half-wave voltage-doubler circuit.

The circuit shown in Figure 3.26 is the half-wave doubler circuit redrawn and labeled such that we can analyze its operation. With S_1 open, we get current i_1 the first time that the input voltage goes negative. Current i_1 will charge C_1 in the direction shown on the circuit diagram in Figure 3.26(a). Once the capacitor is charged to the peak value of the input voltage, there is no discharge path and C_1 holds this peak voltage, as shown in (c). The voltage at point C wrt point D must be the sum of the input voltage and C_1 voltage, as shown in (d). If S_1 is now closed, we have the (approximate) waveform of (d) applied to a rectifier circuit consisting of D_2, C_2, and R_L. With D_2 in the *on* state, C_2 must charge to the peak value of the (d) waveform. The steady-state output waveform across R_L is as shown in (e). The dc output voltage is seen to be nearly twice the peak value of the input and the ripple frequency is the same as the input frequency. Thus, the circuit is accurately described as a half-wave voltage doubler.

A voltage-doubler circuit that will have a ripple frequency twice the input frequency can be built. It is known as a full-wave voltage doubler. The circuit of a full-wave voltage doubler is shown in Figure 3.27. The explanation of this circuit is rather easy, since it consists of diodes used to charge capacitors in opposite directions. The output is taken off the capacitors connected in series such that the dc output voltage is twice the peak input voltage.

The ripple frequency for the circuit of Figure 3.27 is twice the input frequency. If there were no load resistor, the capacitors could never discharge and there would be no ripple voltage. With a load resistor, however, the capacitors can discharge through the load resistance. Since C_1 recharges on the positive input alternation and C_2 recharges on the negative alternation, we have two boosts in the output voltage for each input cycle. Thus the ripple frequency is twice the input frequency and the circuit is called a *full-wave voltage doubler*.

By combining some of the features of the half-wave and full-wave doubler circuits just discussed, we can devise voltage triplers, quadruplers, and higher-voltage multipliers. If we take a half-wave doubler circuit, as shown in Figure 3.25, and add another diode and capacitor, we can obtain a voltage tripler. The tripler circuit is shown in Figure 3.28.

A voltage-quadrupler circuit can be made of two properly connected voltage-doubler circuits. We take two half-wave voltage-doubler circuits and connect them to produce opposite polarity voltages. When we connect the circuits so that the output is taken across the series-connected output capacitors and drive both doublers from the same source, we have a voltage quadrupler. A voltage-

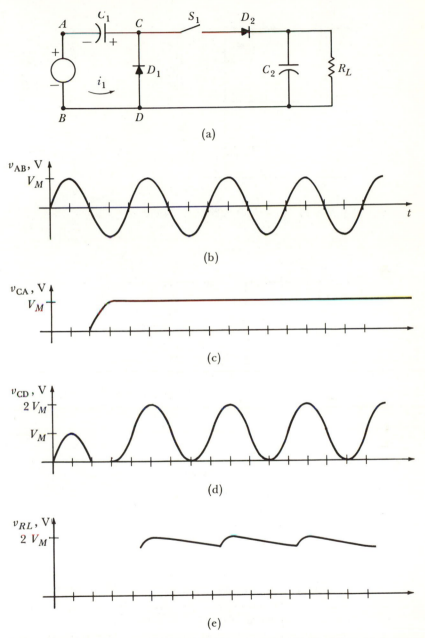

Figure 3.26 Half-wave voltage doubler. (a) Circuit diagram; (b) input voltage (at point A wrt point B) (S_1 open); (c) voltage across capacitor (point C wrt point A) (S_1 open); (d) voltage across D_1 (point C wrt point D) (S_1 open); (e) voltage across R_L (S_1 closed).

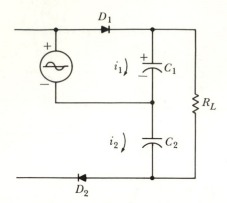

Figure 3.27   Full-wave voltage doubler.

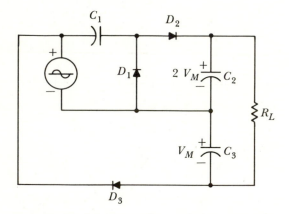

Figure 3.28   Voltage-tripler circuit.

quadrupler circuit connected as previously described is shown in Figure 3.29. Hopefully, the circuit drawing is arranged so that it is apparent that the circuit consists of two doublers connected back to back.

A voltage multiplier can be built to multiply by any number N. The circuit structure shown in Figure 3.30 can be expanded to multiply by the number N needed. The output voltage is taken across the capacitor with the desired multiple of the peak input voltage. The peak inverse voltage rating of each diode must be $2V_M$.

Assume that all diodes and capacitors for the voltage multiplier circuit of Figure 3.30 are ideal. For the ideal case, then, the power delivered by the ac source must be exactly equal to the power dissipated by the load resistance R_L. The source delivers power only

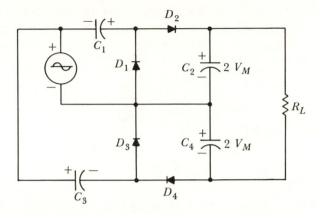

Figure 3.29 Voltage quadrupler.

at the peak of each alternation. The load dissipates power continually since the load voltage is direct current. The power dissipated by the load when the source is not delivering power must come from energy stored in the capacitors. The capacitors deliver energy by releasing some of their stored charge, that is, by supplying discharge current.

Electric power is calculated as the product of voltage and current,

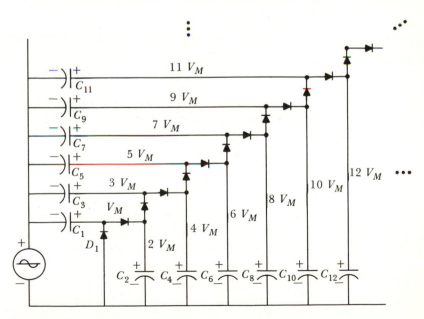

Figure 3.30 Voltage multiplier.

$P = IV$. If we assume that the voltage across the capacitor does not change while delivering the discharge current (obviously not a true assumption, but we will correct for the error later), we find that the following equations are true.

$$P_1 = I_{C1}V_M \qquad \text{(for } C_1\text{)}$$

$$P_2 = I_{C2}2V_M \qquad \text{(for } C_2\text{)}$$

$$P_3 = I_{C3}3V_M \qquad \text{(for } C_3\text{)}$$

$$\vdots$$

$$P_N = I_{CN}NV_M \qquad \text{(for } C_N\text{)}$$

If the power delivered by each capacitor is equal (and it is), then we see that

$$I_{C1}V_P = I_{C2}2V_M$$

$$I_{C1}V_P = I_{C5}5V_M$$

$$\vdots$$

$$I_{C1}V_P = I_{CN}NV_M$$

Solving for the current required from each capacitor, we obtain the following relations:

$$I_{C1} = 2I_{C2}$$

$$I_{C1} = 5I_{C5}$$

$$\vdots$$

$$I_{C1} = NI_{CN}$$

Now that we know the current needed from each capacitor, we can go back and "true up" the obviously false assumption we made in the previous paragraph. Since we know that for any capacitor, $I_C = C(\Delta V_C/\Delta t)$, we know that our previous assumption was at least in partial error. We can minimize the error and at the same time minimize the ripple voltage out of this circuit by making ΔV_C the same for all capacitors. Thus $\Delta V_C = (\Delta t/C)I_C$. The charge and discharge time for all capacitors is the same; thus Δt in all equations is the same. (The discharge time of C_1 is different.) By setting all ΔV_C equal, we can solve for the needed value of C.

$$\Delta V_{C1} = \Delta V_{C2} = \cdots = \Delta V_{C5} = \cdots = \Delta V_{CN}$$

also

$$\frac{\Delta t}{C_1} I_{C1} = \frac{\Delta t}{C_2} I_{C2} = \frac{\Delta t}{C_3} I_{C3} = \cdots = \frac{\Delta t}{C_N} I_{CN}$$

also

$$\frac{I_{C1}}{C_1} = \frac{I_{C2}}{C_2} = \cdots = \frac{I_{C5}}{C_5} = \cdots = \frac{I_{CN}}{C_N}$$

Substituting for the I_C values previously found, we have

$$\frac{I_{C1}}{C_1} = \frac{\frac{1}{2}I_{C1}}{C_2} = \cdots = \frac{\frac{1}{5}I_{C1}}{C_5} = \cdots = \frac{1/N I_{C1}}{C_N}$$

or

$$\frac{1}{C_1} = \frac{1}{2C_2} = \cdots = \frac{1}{5C_5} = \cdots = \frac{1}{NC_N}$$

or

$$C_1 = 2C_2 = \cdots = 5C_5 = \cdots = NC_N$$

We can now find the optimum value for each capacitor where the optimum is based on the value of C_N.

$$C_1 = NC_N$$

$$C_2 = (N - 1)C_N$$

$$C_3 = (N - 2)C_N$$

$$C_i = (N - i + 1)C_N \qquad \text{(The general case where } i \text{ can take on any value between 1 and } N)$$

$$C_N = (N - N + 1)C_N = C_N$$

The voltage multiplier is not commonly used for large multiple values because the circuit has certain drawbacks. The capacitors charge and discharge over a small portion of the input cycle. The current values therefore get very high during the short conduction time. The voltage regulation tends to be poor and the ripple voltage tends to be high for high multiples with significant loads. Obviously the cost of the components can become significant for large multiples.

The circuit can be analyzed quite easily if we consider it in parts. First consider the circuit as containing only the source plus C_1 and D_1. Now note the voltage across the series connection of C_1 and the source. Connect D_2 and C_2 and, assuming no loading of C_1, find the voltage across C_2 and the series connection of C_2 and source. This can be considered as applied to the D_3, C_3 circuit, and so forth.

EXERCISES

QUESTIONS

Q3.1 What is a rectifier circuit?

Q3.2 Why is a rectifier circuit needed?

Q3.3 Why are power-supply filter circuits needed?

Q3.4 In complex power-supply filter circuits, chokes are in series with the load and capacitors are in shunt with the load. Why is this the case?

Q3.5 Why is the rectification efficiency of a full-wave rectifier greater than that of a half-wave rectifier?

Q3.6 Why is the rectification efficiency less than 100% even if we use ideal diodes as rectifiers?

Q3.7 How is a simple shunt-capacitance filter circuit able to reduce the ripple voltage at the load?

Q3.8 What are the advantages and disadvantages of using rectifying voltage-multiplier circuits to obtain high dc voltages?

PROBLEMS

P3.1 Draw the schematic diagram of a transformer-type full-wave rectifier. Show current flow during each alternation.

P3.2 Draw the schematic diagram of a bridge-type full-wave rectifier circuit. Show current during each alternation.

P3.3 The first four terms of a Fourier series for a full-wave rectified sinusoidal waveshape are as follows:

$$f(t) = \frac{20}{\pi} + \frac{40}{3\pi} \cos (2)(377)t - \frac{40}{15\pi} \cos (4)(377)t$$
$$+ \frac{40}{35\pi} \cos (6)(377)t \cdots$$

Plot the sum of these four terms to verify in your own mind that the complete Fourier series would indeed converge to a full-wave rectified waveshape. (*Hint:* Draw each of the waveshapes individually and then find points on the total waveform at various instants of time by adding the values of the individual waveshapes at the same instants of time.)

P3.4 Repeat Problem 3.3 for the following Fourier series terms for a half-wave rectified waveshape.

$$f(t) = \frac{50}{\pi} + 25 \sin 377t - \frac{100}{3\pi} \cos (2)(377)t$$

$$- \frac{100}{15\pi} \cos (4)(377)t \cdots$$

P3.5 A rectifier and filter circuit is shown in Figure 3.31:

(a) Find the peak-to-peak ripple voltage across the load resistor.

(b) Find the dc output voltage.

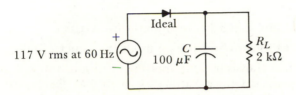

Figure 3.31 See Problems 3.5, 3.6, and 3.7.

P3.6 Repeat Problem 3.5 with a full-wave rectifier.

P3.7 Repeat Problem 3.5 if the input is at 400 Hz.

P3.8 Referring to the circuit of Figure 3.32, what is the minimum value of load resistance that can be used if the peak-to-peak ripple voltage must be 5 V or less?

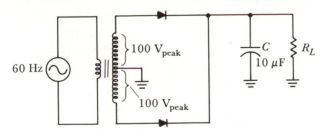

Figure 3.32 See Problems 3.8, 3.9, and 3.10.

P3.9 Repeat Problem 3.8 for a half-wave rectifier.

P3.10 Repeat Problem 3.8 for an input frequency of 1000 Hz.

P3.11 For the circuit shown in Figure 3.33, what value of capacitance C is required in order to keep the peak-to-peak ripple voltage at 10% or less of the peak value of the input-voltage waveform?

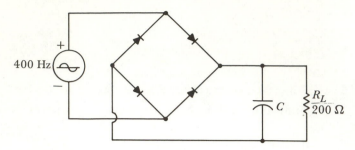

Figure 3.33 See Problems 3.11, 3.12, and 3.13.

P3.12 Repeat Problem 3.11 for a half-wave rectifier.

P3.13 Repeat Problem 3.11 for an input frequency of 60 Hz.

P3.14 A certain transistor amplifier circuit draws 100 mA of current when operated from a battery of about 9 V. Since a 60-Hz ac voltage source of 6.3-V rms is available, it is desired to build a half-wave rectifier circuit with a shunt-capacitance filter circuit to supply the dc voltage required by the amplifier circuit. The peak-to-peak ripple is to be 0.2 V or less.

  (a) Calculate the value of filter capacitance needed.

  (b) Calculate the dc output voltage available.

P3.15 Calculate the ripple to be expected from the filter designed in Problem 3.14 if a full-wave rectifier is used.

P3.16 It is proposed to build a power-supply circuit using the 117-V 60-Hz line voltage as the input. The dc output voltage must be 100 V or more. The peak-to-peak ripple voltage must be 5 V or less. The load resistance is 5000 Ω. Use a full-wave bridge rectifier circuit.

  (a) Calculate the value of filter capacitance needed.

  (b) Calculate the dc output voltage available.

P3.17 Repeat Problem 3.16 with a half-wave rectifier.

P3.18 A 50-V peak-to-peak sinusoidal voltage source drives a half-wave rectifier followed by an L-section filter and a load. The known data is as follows: $L = 1$ H, $C = 60$ μF, $R_L = 200$ Ω, and $f_{in} = 50$ Hz.

  (a) Draw the schematic diagram of the circuit.

  (b) Calculate approximate peak-to-peak ripple voltage across the load.

  (c) Calculate approximate dc output voltage.

P3.19 Repeat Problem 3.18 for $f_{IN} = 400$ Hz.

P3.20 A 20-V peak 60-Hz voltage source is applied to a bridge rectifier and a π-section filter consisting of two 50-μF capacitors and a 1-H choke. The load resistance is 200 Ω.

(a) Draw the circuit diagram.

(b) Calculate the peak-to-peak ripple voltage expected.

P3.21 Repeat Problem 3.20 for a load resistance of 100 Ω.

P3.22 Repeat Problem 3.20 if the filter components are $C_1 =$ 100 μF, $C_2 = 100$ μF, $L = 2$ H, and $R_L = 100$ Ω.

P3.23 A 6.3-V rms 400-Hz signal is applied to a half-wave voltage-doubler circuit.

(a) Draw the circuit diagram.

(b) Draw the output-voltage waveform showing frequency of ripple.

(c) Discuss how applied frequency affects the value of capacitors needed.

CHAPTER 4
SPECIAL-PURPOSE DIODES AND DIODE APPLICATIONS

In addition to the diodes we have discussed thus far, there are several types of diodes that have special characteristics, a partial list of which is as follows:

1. Zener diode (breakdown diode)
2. Gas-regulator diode
3. Field-effect diode
4. Tunnel diode (Esaki diode)
5. Backward diode
6. Hot-carrier diode (Schottky barrier diode)
7. Step-recovery diode (snap diode)
8. Light-emitting diodes
9. Light-sensitive diodes
10. Voltage-variable-capacitance diodes.

In this chapter we will look at the characteristics of several of these diodes and their application in special circuits. Also we will discuss some additional applications of PN junction and vacuum tube diodes.

4.1   ZENER DIODE

The Zener diode is a PN-junction diode whose breakdown voltage in the reverse direction is specified rather accurately by the manufacturer. The manufacturing processes for a PN-junction diode intended for operation as a Zener diode are such that a Zener diode has a very sharp breakdown region.

The *I-V* characteristics shown in Figure 4.1 indicate the comparative sharpness of the "knee" region of the Zener breakdown for an ordinary *PN*-junction diode and a Zener diode. Any *PN* junction has a Zener breakdown in the reverse direction. In the case of an ordinary *PN*-junction diode, the breakdown voltage is made large so that breakdown never occurs in normal operation. The shape of the characteristic curve in the breakdown region is therefore of secondary importance. In the case of the Zener diode, however, the diode is intended to be operating in the breakdown (Zener) region of its characteristic curve. The breakdown region, then, is of primary importance in a Zener diode and is carefully controlled during the manufacturing process.

The specifications that are of greatest importance for a Zener diode are the Zener breakdown voltage and the maximum power the device can safely dissipate. The range of currents a Zener diode can safely handle is a function of the breakdown voltage and the maximum power-dissipation ratings. A Zener diode is normally operating in the breakdown region, so its terminal voltage will be the rated Zener voltage. The voltage, current, and power ratings of a Zener diode are as given by the following equations:

P_Z = rated power dissipation of device

V_Z = rated breakdown voltage of device

I_{max} = maximum current device can safely handle

$P_Z = V_Z I_{max}$

$$I_{max} = \frac{P_Z}{V_Z}$$

The power-dissipation rating of a Zener diode is a function of the ambient temperature in which it is operating, whether or not a heat sink is used, and the size or thermal properties of the heat sink. Zener diodes are available with breakdown voltage ranges from about 3 V to hundreds of volts and with power ratings from about 200 mW to hundreds of watts.

4.2   GAS-FILLED VOLTAGE-REGULATOR TUBES

When electrodes are placed in an atmosphere of an inert gas and a voltage is applied, a useful *I-V* characteristic results. Normally no current conduction is expected in an inert gas. Such is the case until

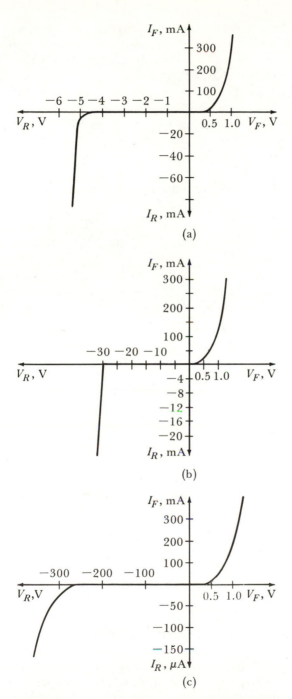

Figure 4.1   *PN*-junction characteristics: (a) typical low-voltage Zener diode; (b) typical high-voltage Zener diode; (c) typical ordinary *PN*-junction diode.

the applied voltage gets large enough to ionize the gas. Ionization in this case is the breaking away of electrons from neutral atoms. Any breaking apart of an atom results in a negative electron and a positive ion. These two charged particles are then free to take part in current conduction. The electron will be accelerated toward the positive electrode and the positive ion will be accelerated toward the negative electrode. The charges moving toward the electrodes will probably collide with other atoms. The energy of collision, acting in addition to the voltage applied between the electrodes, will no doubt cause a further breaking or ionization of the inert gas. The *I-V* characteristics that result from such an experiment are shown in Figure 4.2.

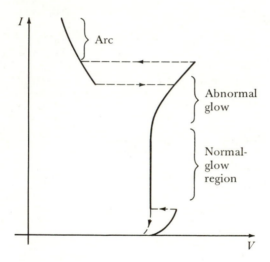

Figure 4.2   Gas-regulator tube (cold-cathode) *I-V* characteristics.

The gas tubes we have been describing have some interesting features we should note. Since neither of the electrodes is heated, the tubes are sometimes called cold-cathode voltage-regulator tubes. We note that there are two paths shown on the *I-V* characteristic as a result of the difference between the breakdown voltage and the regulating voltage. Breakdown voltage must be reached before the tube enters the normal-glow region where the terminal voltage is the tube's regulating voltage. The transition is a step as shown. The path shown in Figure 4.2 for decreasing current is, as pointed out before, different from increasing current. This is easy to explain when we realize that it is easier to maintain ionization, once achieved, than it is to achieve ionization in the first place. Ionization is thus maintained at voltages somewhat less than the breakdown voltage for current in the decreasing direction. A dual path is also noted for operation in the transition region between the abnormal-glow region and the arc discharge region.

For our purposes in this book, we are interested in only one region on the characteristic curve of the cold-cathode gas diode—the normal-glow region. Note the near-constant voltage as current varies in the normal-glow region. Note also that this is the same type of characteristic we had for the Zener diode in the breakdown region. The symbol for, and a possible method of construction of, a gas-regulator tube is shown in Figure 4.3.

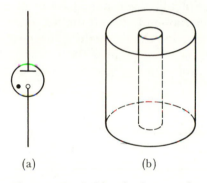

(a)   (b)

Figure 4.3 Cold-cathode gas-voltage-regulator tube. (a) Symbol; (b) possible physical construction.

A sharp point may be part of the cathode structure to help start ionization. When ionized, the tube will glow; thus the familiar term glow tube. Common neon glow tubes (NE 51, for instance) have characteristics similar to those described here. Gas-regulator tubes are most commonly available only in 75-, 90-, 105-, and 150-V ratings. They have thus been largely replaced by Zener diodes, which are available with most any voltage rating from 3 V to hundreds of volts.

4.3 APPLICATIONS OF CONSTANT-VOLTAGE DEVICES

Both the Zener diode and the gas-regulator diode have constant-voltage portions on their respective characteristic curves. For our purposes, we will consider the devices to be equivalent when operating in the region of interest (constant voltage). The most common

application of constant-voltage devices is in voltage-regulator circuits. The circuits can regulate, that is, hold some voltage quantities
nearly constant as other circuit quantities vary. We shall analyze
several of these circuits in detail. As we analyze these circuits, the
reader will achieve an understanding of how and why these circuits
operate.

EXAMPLE 4.1   The circuit shown in Figure 4.4 is designed to hold
the voltage across the load resistance R_L constant at 100 V even as
the input voltage varies over a range of 110 to 130 V. Since we want
to regulate the load voltage at 100 V, it is obvious that the Zener
diode, which is in parallel with the load resistance, must have a 100-
V Zener-breakdown rating. Still to be determined are proper values
for the resistance and power rating of R_1 and the necessary Zener
diode power rating. Note that the Zener diode is installed in the
reverse-bias direction. This is correct, since the constant-voltage
portion of the I-V characteristic is in the reverse direction.

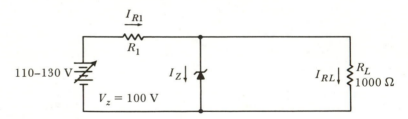

Figure 4.4   Voltage-regulator circuit.

SOLUTION   Some observations of the proposed circuit will be helpful before we try to calculate any numerical values.

1. If the circuit does in fact regulate the load voltage at a constant
value of 100 V while the input voltage may vary between 110 and
130 V, then the voltage across R_1 must vary between 10 and 30 V.

2. If the voltage across R_1 varies between 10 and 30 V while R_1
remains constant (R_1 *is* a constant value), then the current in R_1
must vary.

3. The current in the load resistance R_L must remain constant if R_L
is a constant value and if we do in fact hold the load voltage to a
constant value.

4. If, as the input voltage varies between 110 and 130 V, the current
in R_1 varies (we said above that it did) while the load current in R_L
remains constant, then the current in the Zener diode must vary

because the current in the Zener diode is $I_{R1} - I_{RL}$. The voltage across the Zener diode will remain constant if the circuit is in fact a voltage-regulating circuit.

All these observations can be simultaneously true. The first three are obviously true, based on simple circuit theory. The fourth observation can be true if the Zener diode is operating in the breakdown region below the knee of the characteristic curve. Most Zener diodes are in the (essentially) constant-voltage region below the knee if the reverse current is just a few milliamperes or more. For a specific Zener diode, we would have to look at the characteristic curve on a data sheet to find the minimum current in the constant-voltage region.

The solution of the voltage-regulator-circuit problem of Figure 4.4 involves finding numerical quantities as follows:

1. Find Zener breakdown voltage needed for the Zener diode (already stated to be 100 V).

2. Find the resistance value of R_1.

3. Find the maximum power dissipated by the Zener diode.

4. Find the maximum power dissipated by R_1.

To find the Zener breakdown voltage needed, we merely note in Figure 4.4 that the Zener diode is in parallel with the load resistance. Since we want the load voltage to be 100 V, we must have a 100-V Zener diode. $V_z = 100$ V.

To find R_1, we must notice that the value of R_1 determines the current through itself (I_{R1}) as well as the current through the Zener diode. In fact we note that $I_{R1} = I_z + I_{RL}$.

We know that the output voltage is to be constant at 100 V. Therefore I_{RL} must be constant at

$$I_{RL} = \frac{100 \text{ V}}{1000 \text{ }\Omega} = 0.1 \text{ A} = 100 \text{ mA}$$

Now we know that $I_{R1} = I_z + 100$ mA.

At this point we need to know the minimum current in the Zener diode that will put the operation in the constant voltage region. Let us assume the value to be 5 mA. Therefore

$$I_{R1} \text{ (min)} = 5 \text{ mA} + 100 \text{ mA}$$

$$= 105 \text{ mA}$$

We note from the circuit diagram that the current in R_1 is minimum when the input voltage is a minimum of 110 V. At this time the voltage across R_1 is 10 V. We can calculate R_1 as

$$R_1 \text{ (max)} = \frac{V_{R1}(\min)}{I_{R1}(\min)}$$

$$= \frac{10 \text{ V}}{105 \text{ mA}} = 95.2 \ \Omega$$

We label R_1 as a maximum value because it is based on the minimum current in the Zener diode that will cause the Zener diode to be in the constant-voltage region. Minimum Zener-diode current means minimum R_1 current; and, of course, maximum R_1 resistance corresponds to minimum current.

To find the needed power rating of resistor R_1, we need to know the maximum and minimum current in R_1.

$$I_{R1} \text{ (max)} = \frac{130 - 100}{R_1} = \frac{30}{95.3} = 315 \text{ mA}$$

$$I_{R1} \text{ (min)} = \frac{110 - 100}{R_1} = \frac{10}{95.3} = 105 \text{ mA}$$

$$P_{R1} \text{ (max)} = I_{R1} \text{ (max)} \ V_{R1} \text{ (MAX)} = 315 \text{ mA} \times 30 \text{ V} = 9.45 \text{ W}$$

$$P_{R1} \text{ (min)} = I_{R1} \text{ (min)} \ V_{R1} \text{ (min)} = 105 \text{ mA} \times 10 \text{ V} = 1.05 \text{ W}$$

Thus we see that we will need a 95.3-Ω resistor with a power rating of 9.45 W (or more) if it is to survive for the 130-V input case. The resistor with a 9.45 W power rating will, of course, work fine when dissipating only 1.05 W. We choose the resistor with the 9.45 W (or larger) rating because it can safely handle any input voltage within the specified range.

In order to find the power rating needed for the Zener diode, we will have to determine the maximum and minimum current in the Zener diode.

$$I_Z = I_{R1} - I_{RL}$$

$$I_Z \text{ (max)} = 315 \text{ mA} - 100 \text{ mA}$$

$$= 215 \text{ mA}$$

$$I_Z \text{ (min)} = 105 \text{ mA} - 100 \text{ mA}$$

$$= 5 \text{ mA}$$

$$P_Z \text{ (max)} = I_Z \text{ (max)} \ V_Z = 215 \text{ mA} \times 100 \text{ V}$$

$$= 21.5 \text{ W}$$

$$P_Z \text{ (min)} = I_Z \text{ (min)} \ V_Z = 5 \text{ mA} \times 100 \text{ V}$$

$$= 0.5 \text{ W}$$

We will, of course, choose a Zener diode with a power rating of 21.5 W (or larger) so that the Zener diode can safely operate for any input voltage in the specified range.

A comment is probably due at this point with regard to the values just calculated. The resistance value of 95.3 Ω probably would not be used. A standard value for 5% resistors is 91 Ω. In any event, use the nearest available value of resistance less than 95.3 Ω. If a value of resistance greater than 95.3 Ω is used, the current in the Zener diode will decrease to less than 5 mA. The circuit will not regulate well if this happens. Power resistors are commonly available in ratings of 8, 10, 12, and 20 W. We would choose 10 or 12 W in this case. Zener diodes are available in power ratings of 5, 10, 20, and 50 W. We would have to choose the 50-W rating in this case.

The inclusion of the detailed reasoning for each step undertaken in Example 4.1 tended to make the problem rather long and wordy. Let us, therefore, undertake an example of a similar type but with much less detail.

EXAMPLE 4.2   Design a Zener-diode regulator circuit that will supply a constant 28 V to a 500-Ω load resistance. A voltage source that may vary between 35 and 50 V is available.

SOLUTION
1. First we will draw a circuit diagram. This is shown in Figure 4.5. By inspection of the circuit we see that a 28-V Zener diode is needed if the load voltage is to be 28 V.

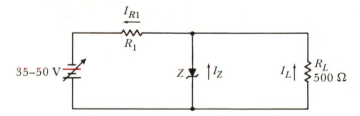

Figure 4.5   Circuit used in Example 4.2.

2. The load current is a constant $I_{RL} = 28$ V/500 $= 56$ mA. By looking at the characteristic curve of a 28-V Zener diode, we can find the minimum current at which the diode is in the constant-voltage region. Let us assume that it is 8 mA in this case. Thus

I_Z (min) $= 8$ mA

The current in R_1 is the sum of load current and Zener current.

I_{R1} (min) = $I_L + I_Z$ (min)

$\qquad$ = 56 mA + 8 mA = 64 mA

3. I_{R1} (min) occurs when the input voltage is minimum. Thus we have

V_{in} (min) = 35 V

V_{R1} (min) = (35 − 28) V = 7 V

$$R_1 = \frac{V}{I} = \frac{7 \text{ V}}{64 \text{ mA}} = 109 \ \Omega$$

NOTE The R_1 value just calculated is usually called the maximum value of R_1. The circuit will operate with a smaller value of resistance. A large resistance is desirable to minimize the power dissipation in R_1 and the Zener diode. The value of R_1 just calculated is the maximum value we can use and still keep the Zener diode in the constant-voltage region (that is, at 8 mA or more).

R_1 (max) = 109 Ω

4. The maximum voltage across R_1 occurs when the input voltage is highest. Thus we have

V_{R1} (max) = 50 − 28 = 22 V

$$I_{R1} \text{ (max)} = \frac{V_{R1} \text{ (max)}}{R_1} = \frac{22}{109} = 0.202 \text{ A}$$

P_{R1} (max) = V_{R1} (max) I_{R1} (max)

$\qquad$ = 22 V × 202 mA = 4.44 W

5. The maximum current in the Zener diode occurs when input voltage is maximum. (I_{R1} is also maximum.) Thus we have

$\qquad I_Z = I_{R1} - I_L$

I_Z (max) = I_{R1} (max) − I_L

I_Z (max) = 202 mA − 55 mA

$\qquad$ = 147 mA

P_Z (max) = I_Z (max) V_Z

$\qquad$ = 147 mA × 28 V

$\qquad$ = 4.12 W

Collecting all the calculated data, we see that we need a 109-Ω 4.44-W resistor and a 28-V, 4.12-W, Zener diode. A good design with a little safety factor built in would probably use a resistance of 100 Ω and a 5-W resistor. The Zener diode chosen would probably be a 5-W, 28-V unit.

This type of regulating circuit is in common use. A typical example would be the electronic circuits in aircraft. In many aircraft, a dc voltage is available that fluctuates quite a bit in magnitude. Since many of the electronic circuits aboard an aircraft require a constant dc voltage, a Zener-diode regulator such as the one just described is often used.

Another type of regulating circuit commonly required will be considered next. The need for the circuit arises when a constant voltage across a changing load is necessary. When we speak of changing load we refer to a changing load current. The circuit of Figure 4.6 will in fact regulate the voltage across a changing load. Note that this circuit is basically the same as the one shown in Figure 4.4.

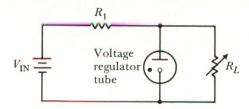

Figure 4.6   Voltage-regulator circuit (variable-load case).

EXAMPLE 4.3   A circuit is needed to regulate the voltage across a changing load. Design such a circuit. The following requirements are imposed on the circuit:

1. Load voltage $V_L = 75$ V.
2. Input voltage $V_{IN} = 90$ V.
3. Load current range, 2 to 30 mA.

SOLUTION
1. First we must draw the circuit diagram of the proposed circuit. Such a diagram is given in Figure 4.7. Note that we are proposing a gas voltage-regulator tube.

2. By inspection we see that a 75-V constant-voltage device is needed. Since a 75-V cold-cathode regulating tube (OA3) is commonly available, let us see if it will satisfy our design requirements. A check of the ratings for an OA3 indicates that the normal-glow region is from $I_Z = 5$ mA to $I_Z = 40$ mA. If we can design the circuit so that I_Z is always in the 5 to 40 mA range, then we can in fact use an OA3 tube.

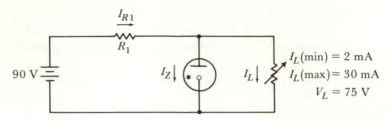

Figure 4.7   Circuit used in solving Example 4.3.

3. Since V_{IN} = 90 V and V_L = 75 V, the voltage across R_1 must be constant. V_{R1} = 90 − 75 = 15 V

4. Since the voltage across R_1 is constant, the current in R_1 must be constant. I_{R1} must also be the sum of I_Z and I_L.

I_{R1} = constant

$I_{R1} = I_Z + I_L$

Solving for I_Z, we have

$I_Z = I_{R1} - I_L$

$I_Z\,(\text{max}) = I_{R1} - I_L\,(\text{min})$

$\qquad\quad = I_{R1} - 2\ \text{mA}$

$I_Z\,(\text{min}) = I_{R1} - I_L\,(\text{max})$

$\qquad\quad = I_{R1} - 30\ \text{mA}$

Some careful consideration of the circuit and a look at the I_Z equation should convince us that the regulator-tube current varies inversely as compared with the load current. It is of course desirable to keep I_Z at a minimum value to conserve power. The minimum current in the normal-glow region is 5 mA. Thus

$I_Z\,(\text{min}) = I_{R1} - 30\ \text{mA}$

$\quad 5\ \text{mA} = I_{R1} - 30\ \text{mA}$

$\qquad I_{R1} = 35\ \text{mA}$

Now we can solve for $I_Z\,(\text{max})$:

$I_Z\,(\text{max}) = I_{R1} - 2\ \text{mA}$

$\qquad\quad = 35\ \text{mA} - 2\ \text{mA}$

$\qquad\quad = 33\ \text{mA}$

Since $I_Z\,(\text{max})$ is in the normal-glow region, we can continue the design knowing that the OA3 tube is satisfactory. If $I_Z\,(\text{max})$ had been larger than could be handled by a glow tube, we could have chosen a suitable Zener diode.

$$R_1 \,(\text{max}) = \frac{V_{R1}}{I_{R1}} = \frac{15 \text{ V}}{35 \text{ mA}} = 429 \text{ }\Omega$$

$$P_{R1} = I_{R1}V_{R1} = 35 \text{ mA} \times 15 \text{ V} = 0.525 \text{ W}$$

A good design choice in this case would probably be a standard value resistor of 430 Ω with a 1-W rating. A Zener diode could replace the gas-regulator tube if desired.

An example of the changing-load phenomenon would be a digital system. (Some might call any digital system a digital computer.) Most digital systems contain a great number of gating circuits each of which are either entirely *on* or entirely *off*. The inputs to the system determine how many gates are *on* and how many are *off*. Thus it is possible (and common) for the small-signal input lines to a digital system to cause the system to draw widely varying amounts of current from its power supply.

The two cases of voltage regulators just discussed are common. Occasionally a circumstance arises where it is desirable to regulate a voltage across a load when the load resistance is changing and also the input-supply voltage is changing. The basic circuit used to accomplish this feat is the same as that of Figures 4.4 and 4.6. It is redrawn in Figure 4.8, showing the variable input and the variable load.

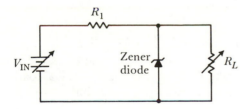

Figure 4.8 Voltage-regulator circuit (variable-input and variable-load case).

EXAMPLE 4.4   Design a Zener-diode voltage-regulator circuit that will satisfy the following requirements:

1. $V_L = 12$ V.

2. $V_{\text{in}} = 20$ to 35 V

3. $I_L = 100$ mA to 1 A

Since we have two quantities changing in this circuit, we must be very careful to consider all possibilities.

SOLUTION

1. The first step in the solution, as usual, is to draw the circuit diagram. This has been done as shown in Figure 4.9.

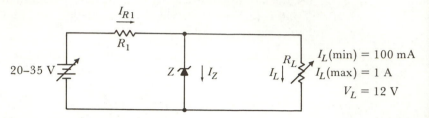

Figure 4.9   Circuit used in solution of Example 4.4.

2. By inspection of the circuit diagram, we can determine the voltage across R_1.

V_{R1} (max) = 23 V

V_{R1} (min) = 8 V

3. By inspection we can determine the Zener voltage needed.

$V_Z = 12$ V

By consulting the data sheet of the manufacturer of the Zener diode we intend to use, we can determine the minimum Zener-diode current in the constant-voltage region. Let us assume

I_Z (min) = 8 mA

4. We have just determined that the very minimum current in the Zener diode should be 8 mA. It should be noted that the current in the Zener diode is a function of input voltage as well as load current. By inspection we see that $I_{R1} = I_Z + I_L$ and that $I_Z = I_{R1} - I_L$.

5. Since I_Z is a function of both I_{R1} and I_L let us, for convenience, consider the functional relationship between I_Z and I_{R1} (as though I_L were constant) and then consider the functional relationship between I_Z and I_L (considering I_{R1} as constant).

(a) If I_L were constant, we would observe the following:

I_Z (min) occurs when I_{R1} (min) occurs

I_{R1} (min) occurs when V_{IN} (min) occurs

∴ I_Z (min) occurs when V_{IN} (min) occurs

(b) If V_{in} (and thus I_{R1}) were considered constant, we would observe that I_Z (min) occurs when I_L (max) occurs. If it is not obvious that this statement is true, go back and reread and rework Example 4.2. Combining the two considerations made separately in the preceding, we can conclude the following:

I_Z (min, min) occurs when V_{in} (min) occurs *and* when I_L (max) simultaneously occurs.

I_Z (min, min) = I_{R1} (min) − I_L (max).

8 mA = I_{R1} (min) − 1A.

I_{R1} (min) = 8 mA + 1000 mA = 1008 mA.

6. Since we know that I_{R1} (min) occurs when V_{IN} (min) occurs, we can calculate the value of resistance R_1.

$$R_1 \text{ (max)} = \frac{V_{IN} \text{ (min)} - V_L}{I_{R1} \text{ (min)}} = \frac{V_{R1} \text{ (min)}}{I_{R1} \text{ (min)}}$$

$$R_1 \text{ (max)} = \frac{8 \text{ V}}{1008 \text{ mA}} = 7.94 \text{ } \Omega$$

The R_1 value just calculated is labeled maximum because it is the largest value we can use that will ensure that I_Z will remain in the constant-voltage region with a current of 8 mA or more.

7. By inspection we observe that I_{R1} (max) will occur when V_{in} (max) occurs. Thus

$$I_{R1} \text{ (max)} = \frac{V_{R1} \text{ (max)}}{R_1} = \frac{23}{7.94} = 2.9 \text{ A}$$

P_{R1} (max) = I_{R1} (max) V_{R1} (max) = 2.9 × 23 = 66.8 W

8. By a process similar to that undertaken in step 4, we can conclude that I_Z (max, max) occurs when V_{in} (max) occurs *and* when I_L (min) simultaneously occurs.

I_Z (max, max) = I_{R1} (max) − I_L (min)

I_Z (max, max) = 2.9 A − 100 mA

$$= 2.8 \text{ A}$$

$$P_Z = I_Z \text{ (max, max)} V_Z = 2.8 \times 12$$

$$= 33.6 \text{ W}$$

Collecting all the data calculated for this example problem, we have the following values:

$V_Z = 12 \text{ V}$

$P_Z = 33.6 \text{ W}$

$R_1 = 7.95 \text{ } \Omega$

$P_{R1} = 66.8 \text{ W}$

If it is determined that we could allow I_Z to be just slightly less than 8 mA in the worst case, then we would choose $R_1 = 8\ \Omega$. If this is not the case, then we would choose R_1 to have a resistance slightly less than (or equal to) 7.95 Ω and a power rating greater than 6.68 W. The nearest standard value would be 7.5 Ω and 50 W. The Zener diode would obviously need a 12-V rating and its power rating should be 50 W.

4.4   FIELD-EFFECT DIODE

A diode whose I-V characteristic can be considered as being inverse or opposite to that of the Zener diode is the field-effect diode. The specifications for a family of field-effect diodes is shown in Figure 4.10. Take particular note of the I-V characteristics and the schematic symbol.

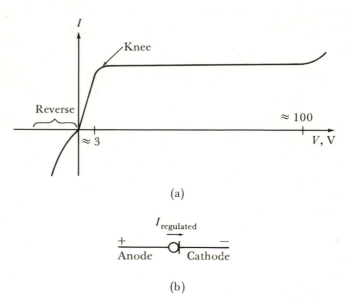

Figure 4.10   Field-effect diode. (a) I-V characteristics; (b) symbol.

Field-effect diodes (that is, current-regulating diodes) are commercially available in a more limited range of values than is the case for Zener diodes (that is, voltage-regulating diodes). Field-effect diodes became available several years later than Zener diodes and are less widely used and more expensive than Zener diodes.

There are several commonly used circuit functions that seem to be natural applications for current-regulator diodes. Among these applications are the ramp-generator circuits and the staircase-generator circuits. Let us discuss two example circuits to gain an understanding of possible applications.

EXAMPLE 4.5   A circuit is needed to generate a voltage sawtooth waveform. (This is a commonly required waveform. Every oscilloscope and television receiver has some sort of sawtooth or ramp-generator circuit.) The waveform might look like one of those in Figure 4.11(a). Note that the sawtooth waveform is made up of linear voltage ramps.

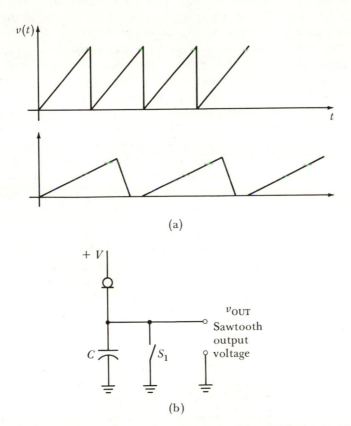

(a)

(b)

Figure 4.11   (a) Sawtooth voltage waveforms of Example 4.5. (b) A proposed sawtooth, or ramp-generating, circuit.

SOLUTION   A possible method of generating a voltage ramp makes use of the *I-V* relationships of a capacitance.

$$i_C = C \frac{dv_c}{dt}$$

The relationship for i_C can be approximated by

$$i_C = C \frac{\Delta v_c}{\Delta t}$$

where the Δ (delta) means a change in value.

A careful look at the equation for the current in a capacitance should convince us that if i_C is a constant, then $\Delta v/\Delta t$ must be a constant value (assuming the value of the capacitance C is constant). The constant value of $\Delta v/\Delta t$ is of special interest to us at this time, since it exactly describes a linear voltage ramp.

If we can somehow connect a current-regulator diode to supply a constant current into a capacitor, the voltage across that capacitor must be a linear ramp. The circuit shown in Figure 4.11(b) is proposed as a sawtooth, or ramp-generating, circuit. When switch S_1 is open, the constant current in the diode is also I_c, which results in a ramp of voltage across the capacitor. When the output voltage (that is, V_c) is of proper magnitude, switch S_1 is closed to discharge C. Switch S_1 may be closed only instantaneously or may be left closed for some finite period of time depending on the output waveshape desired. In most practical cases, switch S_1 would be a transistor used as a switch.

EXAMPLE 4.6   It is desired to build a current staircase-generator circuit. Such circuits are sometimes used in computers or in circuits meant to convert signals from digital (discrete-step) form to analog (continuous) form. It is sufficient at this point to say that staircase-generator circuits are used. A staircase-current waveform is shown in Figure 4.12(a).

SOLUTION   The circuit of Figure 4.12(b) is proposed as a possible current staircase-generator circuit. When all switches are open, current is zero. Current increases in discrete steps as switches S_1 through S_4 are closed in sequence. The load-current waveshape will be as shown in Figure 4.12(c).

If the load is a resistance, then the voltage across the resistance will be a voltage staircase. In a practical situation the mechanical switches would probably be replaced by transistor switches.

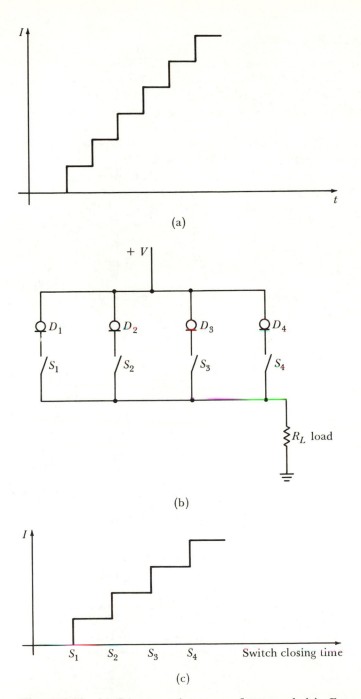

Figure 4.12 (a) Current staircase waveform needed in Example 4.6. (b) Proposed circuit. (c) Current output waveform for the circuit of Figure 4.12(b).

4.5   TUNNEL DIODE

The tunnel diode (Esaki diode) is a modification of a *PN* junction discovered by Dr. Esaki, a Japanese physicist, in 1958. The modifications include increased doping levels among other things. For our purposes, we are interested in the terminal *I-V* characteristics. The tunnel-diode terminal *I-V* characteristics are shown in Figure 4.13. Some of the different symbols used to represent tunnel diodes are shown in Figure 4.13(b). The subscripts *P* and *V* stand for peak and valley, respectively. Peak current, valley current, peak voltage, and valley voltage are defined as shown in the *I-V* characteristics.

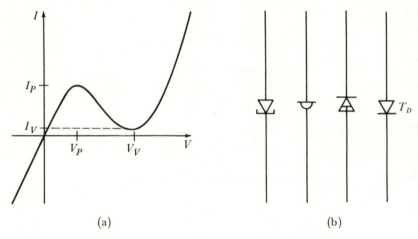

(a)   (b)

Figure 4.13   Tunnel diode: (a) *I-V* characteristics; (b) alternate tunnel-diode symbols.

The *I-V* characteristic of the tunnel diode is obviously nonlinear. At first glance it probably appears unusual *and* not useful. However, upon closer examination of the characteristics we find that the unusual shape represents something very interesting. We note that the characteristic curve represents both positive and negative values of dynamic resistance. The dynamic resistance is negative when the device is operating such that its terminal voltage is between V_P and V_V. If we arbitrarily choose some numbers for the quantities involved, we can calculate the dynamic resistance.

$$I_P = 5 \text{ mA}$$

$$I_V = 0.5 \text{ mA}$$

$V_P = 60$ mV

$V_V = 360$ mV

These values are about what we would expect for an IN3852 tunnel diode.

Let us therefore choose the Δ quantities from peak to valley.

$$r = \frac{\Delta V}{\Delta I} = \frac{V_P - V_V}{I_P - I_V} = \frac{(60 - 360) \text{ mV}}{(5 - 0.5) \text{ mA}}$$

$$= \frac{-300 \text{ mV}}{4.5 \text{ mA}} = -66.7 \ \Omega$$

The negative dynamic resistance means that (in the negative-resistance region) as voltage is increased, current decreases and/or vice versa.

Operation in the negative-resistance region is unstable. If the tunnel diode is placed in a circuit where the possible current is multivalued (this is a very possible situation), the circuit current in the tunnel diode will always go to the positive-resistance region. If the circuit current is somehow made to instantaneously operate in the negative-resistance region, the operating point will quickly go to an operating point where r is positive. A tunnel-diode circuit can be built where the operating point is continually changing from and through the unstable region. A tunnel-diode circuit can therefore act as an oscillator. The frequency of oscillation is controlled by circuit inductors and capacitors. Since the tunnel-diode operation moves through the unstable region so quickly, a tunnel-diode oscillator can be made to generate extremely high frequency signals.

Tunnel diodes have other possible applications. They can be used in amplifier circuits. In computer applications, they can be used in bistable, monostable, and astable flip-flop circuits.

4.6 BACKWARD DIODE

A backward diode is a special adaptation of the tunnel-diode phenomenon. This device is also called a back diode or unitunnel diode. The back diode differs from other tunnel diodes in that it is manufactured so that the peak current and the valley current are very nearly equal and near zero. The characteristic curve of a typical back diode is shown in Figure 4.14.

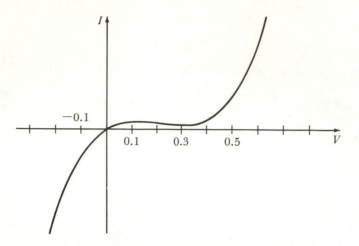

Figure 4.14   Backward-diode characteristics.

Since the characteristics in the forward direction are not much different from those for an ordinary PN-junction diode, it seems logical to assume that the main usefulness of the back diode lies in its reverse-bias characteristics. The reverse characteristics of back diodes are often drawn in the first quadrant of the I-V graph so that comparison with normal PN-junction diodes is easy. In Figure 4.15 we have the back-diode characteristics and the characteristics of PN-junction diodes drawn on the same set of coordinate axes.

Some useful properties of the back diode should be evident in

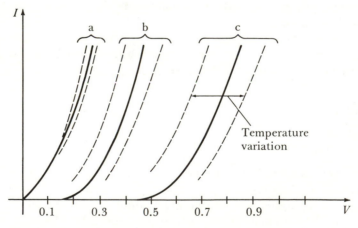

Figure 4.15   Characteristics of back diode and junction diodes: (a) back diode; (b) germanium PN-junction diode; (c) silicon PN-junction diode.

Figure 4.15. The most important property is that the back diode has essentially zero turn-on voltage. The device is also very stable in temperature.

Most applications of the back diode rely on its zero turn-on voltage property. When you want to rectify a signal whose peak voltage is, for example, 0.1 V, a *PN*-junction diode cannot be used because it will never turn on. The back diode can handle the job with ease, however.

4.7   HOT-CARRIER DIODES

A diode can be formed by a semiconductor metal junction. Such diodes are called Schottky barrier diodes or hot-carrier diodes. For our purposes, we will consider them exactly the same as a *PN*-junction diode. Hot-carrier diodes can be built with very short turn-on and turn-off times. In certain high-speed applications, hot-carrier diodes may be advantageous or necessary.

4.8   STEP-RECOVERY DIODES

Another special-application diode is called the step-recovery, or snap, diode. Any *PN* junction has the property that when instantaneously switched from forward bias to reverse bias, current will be present for a very short time in the reverse direction. The step-recovery diode will also conduct for a short time in the reverse direction. The step-recovery diode differs from an ordinary *PN*-junction diode in that the time of conduction in the reverse direction is accurately controlled by the manufacturer. Also the recovery from reverse conduction is unusual in that it is very abrupt, much like a snap action—thus the term *snap diode*. The operation of a step-recovery diode in a circuit as compared with an ordinary *PN*-junction diode is shown in Figure 4.16.

The manufacturer of the step-recovery diode will specify the reverse-conduction time of the diode. By using these diodes, then, it is possible to obtain a current pulse of the desired duration with very fast rise and fall times [see the negative portion of Figure 4.16(c)]. The voltage across the resistor is, of course, the same waveshape as the current. The unwanted portion of the curve could be clipped off with a very-fast junction-diode circuit, perhaps a hot-carrier diode.

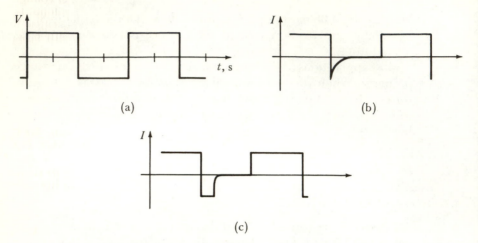

(a)

(b)

(c)

Figure 4.16   Step-recovery diode dynamic reverse-conduction characteristics: (a) applied voltage; (b) current waveform PN-junction diode; (c) current waveform step-recovery diode.

4.9   PHOTODEVICES

In addition to the useful properties we have already studied, the PN junction has some light-sensitive properties. The pictorial representation shown in Figure 4.17 will give some idea about the physical construction involved.

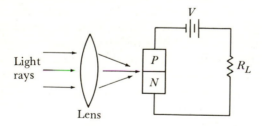

Figure 4.17   Photodiode.

Note that the light is directed toward the junction. Obviously the package containing a photodiode must have some light-conductive window or lens.

PHOTODIODES

The *I-V* characteristics of photodiodes are in some ways similar to the ordinary *PN*-junction diode but must be somewhat different if the device is truly light sensitive. Note in Figure 4.17 that the battery is connected so that the *PN* junction is reverse-biased. Our past experience with *PN* junctions would tell us that the current would be very minimal and near zero. This is in fact the case when no light is present, and the small reverse current that flows is called the *dark current*.

As the applied light intensity is increased, the magnitude of the reverse current increases for a photodiode. The current in the circuit is thus seen to be function of applied voltage as well as light intensity. In order to show these characteristics on a two-dimensional graph, we must use what is known as a family of curves. Under dark conditions, the ordinary *I-V* characteristics of a reverse-biased *PN* junction is the result. Another *I-V* characteristic can be drawn on the graph for some specified light intensity. By repeating the *I-V* characteristics for several different light levels, we end up with a family of curves as shown in Figure 4.18. The different illumination levels are indicated as dark, L_1, L_2, L_3, and L_4. The units for illumination may be foot-candles or some other units depending on the system of units employed.

Hopefully it is obvious at this point that the photo diode is intended to be reverse-biased so that the magnitude of current in the device will be a function of illumination. The photodiode is there-

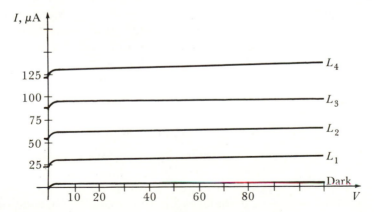

Figure 4.18 Photodiode characteristics.

fore not a bilateral device. Its characteristics are quite different when operated in the forward-bias mode. When operated in the reverse-bias mode, the photodiode could be described as photoresistive (or photoconductive), since its resistance (or conductance) is a function of the level of illumination.

BILATERAL PHOTORESISTIVE DEVICE

Another type of photoresistive (photoconductive) device is bilateral. It consists of a thin layer of photosensitive semiconductor deposited on an insulator with metallic conductive leads at each end of the deposited strip. A simple representation of the construction is shown in Figure 4.19.

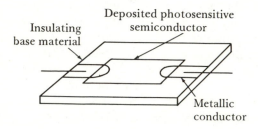

Figure 4.19   Photoresistive device.

The photosensitive semiconductor materials used are quite high in resistivity with no illumination. The light energy is sufficient to enable some of the covalently bonded electrons to break their covalent bonds and thus be available for conduction. Such devices are capable of very large changes in resistance as the illumination varies, but the functional relationship is very nonlinear. Cadmium sulfide (CdS) is a commonly used photosensitive semiconductor. This type of photo device is not based upon a PN-junction phenomenon but is included here for comparison purposes.

PHOTOVOLTAIC DEVICES (SOLAR BATTERIES)

A close inspection of the photodiode characteristics will show a very interesting property. Even if the diode is shorted so that the terminal voltage is zero, a current results. The magnitude of the current is a function of the magnitude of the illumination. The photodiode, then, can be considered as a voltage source or battery. These photovoltaic devices produce small voltages (0.5 V or less) and current up to several milliamperes. To get increased voltage, solar batteries are connected in series. The open-circuit voltage versus illumination and the short-circuit current versus illumination are shown on the same graph in Figure 4.20.

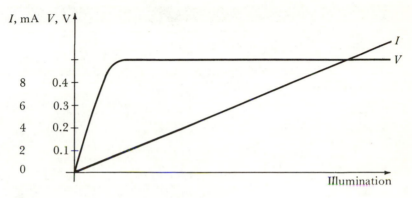

Figure 4.20   Solar-battery characteristics.

Some of the more obvious applications of photoresistive devices and photovoltaic devices are

1. Burglar-alarm systems

2. Safety device on passenger-elevator doors.

3. Card or tape-reader devices to determine presence or absence of holes in data-processing systems.

They are used extensively in the industrial situation to count units coming down a production line. Many inspection devices and methods make use of photosensitive components.

LIGHT-EMITTING DIODES

Another photodevice is the light-emitting diode, or LED. It is a device that will emit light through a window or lens when there is junction current. At the present state of development, the magnitude of the emitted light is rather small, which indicates poor efficiency in conversion of electric energy into light energy. The most efficient units produce a red light. Much of the usefulness of these units is owing to the small size of the units.

A common application of light-emitting diodes is in alphanumeric readouts. If a number of LEDs are arranged properly, almost any

Monsanto | .27" RED SEVEN SEGMENT DISPLAY | MAN1 MAN1A

PRODUCT DESCRIPTION

The MAN1 is a seven segment diffused planar GaAsP light emitting diode array. It is mounted on a dual in-line 14 pin substrate and then encapsulated in clear epoxy for protection. It is capable of displaying all digits and nine distinct letters. The MAN1A has identical specifications, but is encapsulated in high contrast red epoxy.

PACKAGE DIMENSIONS

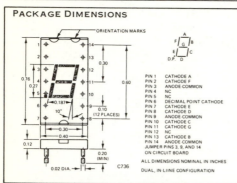

PIN 1   CATHODE A
PIN 2   CATHODE F
PIN 3   ANODE-COMMON
PIN 4   NC
PIN 5   NC
PIN 6   DECIMAL POINT CATHODE
PIN 7   CATHODE E
PIN 8   CATHODE D
PIN 9   ANODE-COMMON
PIN 10   CATHODE C
PIN 11   CATHODE G
PIN 12   NC
PIN 13   CATHODE B
PIN 14   ANODE COMMON
JUMPER PINS 3, 9, AND 14
ON CIRCUIT BOARD

ALL DIMENSIONS NOMINAL IN INCHES

DUAL, IN-LIINE CONFIGURATION

FEATURES

- High brightness . . . Typically 350 ft-L @ 20 mA
- Single plane, wide angle viewing . . . 150°
- Unobstructed emitting surface
- Standard 14 pin dual-in-line package configuration
- Long operating life . . . solid state reliability
- Shock resistant
- Operates with IC voltage requirements
- Small size; offering unique styling advantages
- All numbers plus 9 distinct letters
- Usable for wide viewing angle requirements
- Usable in vibrating environment, impervious to vibration
- Directly compatible with integrated circuits

The MAN1 is for industrial and military applications such as:

- Digital readout displays
- Cockpit readout displays

ABSOLUTE MAXIMUM RATINGS

Power dissipation @ 25°C ambient . . . . 750 mW
Derate linearly from 25°C . . . . 10 mW/°C
Storage and operating temp . . . . -55°C to 100°C
Continuous forward current
    Total . . . . 240 mA
    Per segment . . . . 30 mA
    Decimal point . . . . 30 mA
Reverse Voltage
    Per segment . . . . 6.0 volts
    Decimal point . . . . 3.0 volts

ELECTRO-OPTICAL CHARACTERISTICS

(25°C Ambient Temperature Unless Otherwise Specified)

CHARACTERISTICS	MIN.	TYP.	MAX.	UNITS	TEST CONDITIONS
Brightness (note 1)					
Segment	100	350		ft-L	I_F=20 mA, λ=650 nm
Decimal point	100	350		ft-L	I_F=20 mA, λ=650 nm
Peak emission wave length	630		700	nm	
Spectral line half width		20		nm	
Forward voltage					
Segment		3.4	4.0	V	I_F=20 mA
Decimal point		1.6	2.0	V	I_F=20 mA
Dynamic resistance					
Segment		11		Ω	I_F=20 mA
Decimal point		5.5		Ω	I_F=20 mA
Capacitance					
Segment		80		pF	V=0
Decimal point		135		pF	V=0
Reverse Current					
Segment			100	μA	V_R=6.0 volts
Decimal point			100	μA	V_R=3.0 volts

Figure 4.21 Seven-segment numeric display using light-emitting diodes (LED). (Courtesy of Monsanto Company).

alphanumeric symbol can be produced by illuminating only selected units. Figure 4.21 shows a manufacturer's data sheet for a seven-segment numerical readout using LEDs as the light source.

4.10   VOLTAGE-VARIABLE CAPACITOR

PN-junction diodes can be used as voltage-variable capacitors. When used in this mode they are called *varactors*, or voltage-variable capacitors.

In order to explain the variable-capacitance feature of a *PN*-junction diode, we will need to review some basic facts about junctions. Figure 4.22 shows a reverse-biased *PN*-junction diode. The *P*- and *N*-type materials are semiconducting and the depletion region is nonconducting. Thus we have conducting materials separated by an insulator. This is of course a reasonable approximation of a parallel-plate capacitor. The drawing of Figure 4.22 shows us that a reverse-biased *PN* junction will have some capacitance as measured between its terminals.

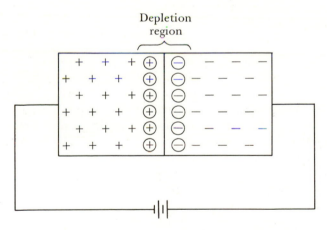

Figure 4.22   Reverse-biased *PN*-junction diode.

The junction capacitance of a *PN*-junction diode is interesting and useful because of its voltage-variable feature. As we learned earlier, the width of the depletion region of a junction diode is dependent on the amount of reverse voltage applied. The width of the depletion region is effectively the distance between the plates of a parallel-plate capacitor. As the reverse-bias voltage increases, the distance between the plates of the capacitor increases (that is, the width of the depletion region increases) and capacitance decreases; thus it is called a voltage-variable capacitor.

The voltage-variable capacitor can be used in many tuning applications. A resonant circuit can be tuned (that is, its frequency can be adjusted) with a varactor, as shown in Figure 4.23. The components that determine the resonant frequency are L_1 and C_1.

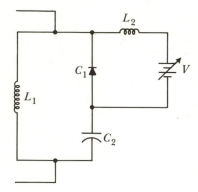

Figure 4.23 A voltage-variable capacitor (C_1) in a tuned-circuit application.

Let us consider why the components L_2 and C_2 are included in the circuit of Figure 4.23. The capacitance C_2 is needed so that the battery, that is, the capacitance-varying dc voltage, does not see a dc short circuit in the series connection of L_1 and L_2. The inductance L_2 is needed so that L_1 does not see the ac short circuit of the battery in parallel with it.

The ac equivalent of the circuit shown in Figure 4.23 is a simple parallel connection of L_1 and C_1. This is the case if both L_2 and C_2 are made very large. With L_2 and C_2 large, X_{L2} approaches infinity and X_{C2} approaches zero. Inductance L_1 is in parallel with the series connection of X_{C1} and X_{C2}. Since we have specified that X_{C2} is near zero, the sum of X_{C1} and X_{C2} is very nearly equal to X_{C1}. Thus L_1 sees only C_1 or X_{C1}. Since L_2 is large, the series connection of L_2 and the battery is essentially an open circuit to signals. With L_2 an open circuit and C_2 a short circuit, L_1 is effectively in parallel only with C_1 (in terms of ac).

4.11   OTHER DIODE APPLICATIONS

METER PROTECTION

Diodes may be used to protect meter movements from current overload damage. A 1-mA dc-meter movement typically has 50 Ω of internal resistance. Suppose that we want to protect such a movement by using a germanium *PN*-junction diode. Assume that the turn-on voltage for the diode is 0.3 V and we place the diode directly across the meter, as shown in Figure 4.24. Let us calculate the

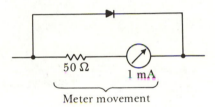

Figure 4.24   Meter-protection diode circuit.

current through the meter that will result in 0.3 V across the meter movement.

$$I = \frac{V}{R} = \frac{0.3 \text{ V}}{50 \text{ Ω}} = 6 \text{ mA}$$

Obviously this diode does not start bypassing current around the meter until the meter is overloaded by a factor of 6. If the application will allow some added series resistance, we can design for a bypassing action at, say, 2 mA. We can cause this to happen by adding enough resistance R so that the voltage across the diode is 0.3 V when the current through the ammeter is 2 mA.

$$V = IR_{\text{TOTAL}} = I(R + 50)$$

$$R + 50 = \frac{V}{I}$$

$$R = \frac{V}{I} - 50$$

$$R = \frac{0.3 \text{ V}}{2 \text{ mA}} - 50$$

$$R = 100 \text{ Ω}$$

The circuit shown in Figure 4.25 will start to bypass some current around the meter movement when the line current exceeds 2 mA if $R = 100 \ \Omega$. If we knew the equivalent resistance of the diode, we could calculate the amount of bypassed current for any line current.

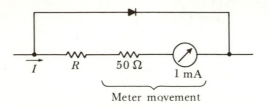

Figure 4.25   Meter-protection circuit.

METER RECTIFIERS

Many times diode rectifiers are part of voltmeters intended to measure ac voltage. Since most meter movements are dc-operated, an ac-to-dc conversion is necessary in order to apply a dc movement to the application of measuring ac voltages. The circuit of Figure 4.26 shows both dc and ac voltmeter circuits.

If the ac voltmeters shown in Figure 4.26(b) and (c) are intended for use in low-voltage circuits, the diodes to be used should have low turn-on voltages.

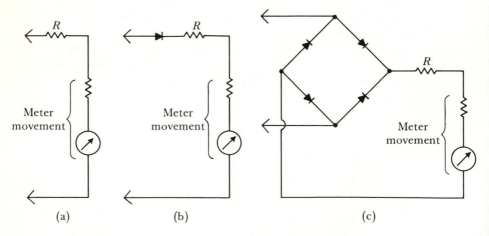

Figure 4.26   Voltmeter circuits: (a) dc voltmeter; (b) ac voltmeter with half-wave rectifier; (c) ac voltmeter with full-wave rectifier.

CALIBRATION OF VOLTMETERS USING RECTIFIERS

Most meter movements are basically current-operated devices. When used in conjunction with a series resistor to measure voltage, the scale is usually calibrated in volts. For instance, the 1-mA position may be marked as 100 V. A 100-V full-scale voltmeter using a 1-mA 50-Ω movement would appear as in Figure 4.27.

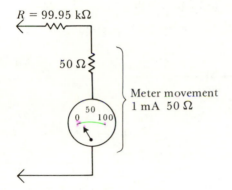

$R = 99.95$ kΩ

50 Ω

Meter movement
1 mA 50 Ω

Figure 4.27   Voltmeter circuit.

Resistance R is chosen so that when the voltage between the probes is 100 V, the current is 1 mA, and the needle will point to 100 V on the scale. The R value is calculated as follows:

$$R + 50 = \frac{100}{1 \text{ mA}} = 100 \text{ k}\Omega$$

$$R = 100 \text{ k}\Omega - 50 \text{ }\Omega = 99.95 \text{ k}\Omega$$

We can define the scale factor for this circuit as the relationship between the current in the meter and the voltage value printed on the face of the meter.

Voltage reading = IS where S is the scale factor. For the circuit shown in Figure 4.27, the scale factor is as follows:

$$V = IS$$

$$100 = (1 \text{ mA}) S$$

$$S = \frac{100 \text{ V}}{1 \text{ mA}} = 100 \text{ kV/A}$$

Voltmeters meant to be used for ac-voltage measurements must have a rectifier diode. In addition, they will require a different scale factor or a different series resistor than that for a similar dc unit. To show that this is so, we will use the meter circuit of Figure 4.27, add a rectifier diode, and show that the meter reading will be the wrong value for the given resistance R and scale factor S.

EXAMPLE 4.7   It is desired to add only a diode (ideal) to the circuit of Figure 4.27 and by so doing have a meter to measure correctly the rms value of a sinusoidal waveshape. The proposed circuit is shown in Figure 4.28(a).

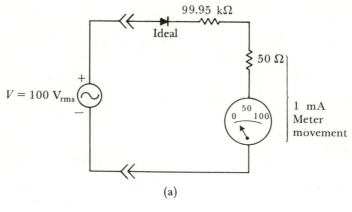

(a)

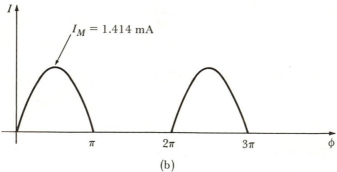

(b)

Figure 4.28   (a) Circuit for Example 4.7. (b) Current waveform through the meter.

SOLUTION   The current waveform in the meter movement is a half-wave rectified sine wave. Since the movement responds to the average current, we must find the average value of the waveform. The peak value of the waveform is

$$I_M = \frac{V_M}{R + 50} = \frac{100\ (\sqrt{2})}{100\ k\Omega} = 1.414\ \text{mA}$$

The current waveform is shown in Figure 4.28(b). The average current is calculated as

$$I_{av} = \frac{\text{area}}{\text{base}} = \frac{2I_M}{2\pi} = \frac{1.414\ \text{mA}}{\pi} = 0.45\ \text{mA}$$

The meter reads

$$V = (I_{av})(\text{scale factor})$$

$$= (0.45\ \text{mA})(100\ \text{kV/A})$$

$$= 45\ \text{V}$$

We have just proved that when we attempt to read the rms magnitude of a sinusoidal voltage with a dc meter with a rectifier added, we read less than one-half the true rms value. Thus, to get a correct reading rms voltmeter, we must paint new numbers on the face of the meter movement or reduce the value of R so that I_{av} is 1 mA.

One more example should be sufficient to teach the principles involved in calibrating ac-voltmeter circuits.

EXAMPLE 4.8   We have available a 200-μA meter movement with internal resistance of 200 Ω. We want to build a voltmeter circuit that will read the peak value of a square-wave signal. The full-scale reading is to be 50 V. The circuit is to consist of a standard voltmeter circuit with a rectifier diode (ideal) added. Calculate the value of series resistance needed.

SOLUTION
1. First we must paint numbers on the face of the movement so that the 200-μA deflection will now read 50 V peak.

$$V = I_{av} S; \qquad S = \frac{V}{I} = \frac{50\ \text{V}}{200\ \mu\text{A}} = 250\ k\Omega \quad \text{or } 250\ \text{kV/A}$$

2. The circuit and waveforms are drawn as shown in Figure 4.29.

3.

$$I_{av} = \frac{\text{area}}{\text{base}} = \frac{I_M \pi}{2\pi} = \frac{I_M}{2} = \frac{V_M}{2(R + 200)}$$

$$= \frac{25}{R + 200}$$

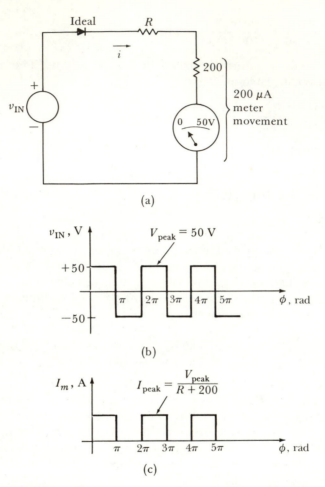

Figure 4.29 Circuit and waveforms for Example 4.8. (a) Circuit diagram; (b) input-voltage waveform for full-scale deflection; (c) meter current waveform.

4. Since we have 50 V on the movement face, we want an average current of 200 μA when a 50-V-peak square wave is applied.

$$I_{av} = \frac{25}{R + 200} = 200 \ \mu A$$

$$25 = (200 \ \mu A)(R + 200)$$

$$R + 200 = \frac{25}{200 \ \mu A} = 125 \ k\Omega$$

$$R = 125 \ k\Omega - 200$$

$$= 124.8 \ k\Omega$$

We now have a meter that will accurately read the peak voltage of square-wave signals.

A dc ammeter may be used in an ac voltmeter circuit by using a rectifying diode as in the Examples 4.7 and 4.8. By using techniques similar to the example problems, we can design the circuit to read correctly for any known waveshape. We can also design the voltmeter circuit to read in peak, peak-to-peak, or rms voltage.

FREE-WHEELING DIODES

In many pieces of equipment we find it necessary, or at least convenient, to suddenly make or break a circuit with a switch contact. We may experience considerable difficulty if we suddenly open a current-carrying inductive circuit. Consider the circuit of Figure 4.30(a). It is a simple inductive circuit containing a switch.

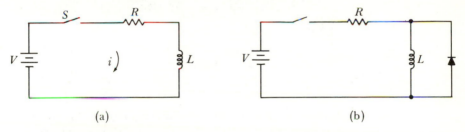

(a)   (b)

Figure 4.30   Switching in an inductive circuit: (a) without free-wheeling diode; (b) with free-wheeling diode.

Let us look carefully at the operation of Figure 4.30(a) during switching. When S is closed, we expect the current to increase exponentially, as shown in Figure 4.31. When switch S opens, the entire circuit is open and current is zero. We normally assume that the switch opens instantly and thus the current stops immediately.

In order to evaluate the voltages around the loop during the

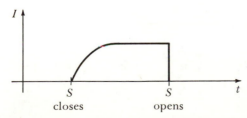

Figure 4.31   Current waveform for circuit of Figure 4.30(a) (idealized).

switching operations, let us consider the *I-V* characteristics of an inductance.

$$v_L = L \frac{di_L}{dt}$$

When S closes, we see that di/dt is some finite value. In fact, since V_R is zero at the first instant, we see that

$$v_L = V$$

Thus

$$v_L = V = L \frac{di}{dt}$$

$$\frac{di}{dt} = \frac{V}{L}$$

Of greater interest is the time when S opens. In Figure 4.31 we see that the current changes to zero instantly, which makes $di/dt = -\infty$. The voltage across L is then

$$v_L = L \frac{di}{dt} = L(-\infty)$$

$$= -\infty$$

These calculations show what would happen if the current through an inductance could be stopped instantly. If we attempt to do this, we will get a high voltage, which will cause arcing across the switch contacts to maintain the current flowing for a short time. Obviously the arcing can cause damage to the switch and perhaps other components in a more complex circuit.

In many cases, a solution to the arcing problem associated with opening a switch in an inductive circuit can be found. The free-wheeling diode circuit is one such solution. In Figure 4.32 we see that i_2 is the current when S is open. The current i_2 decays to zero with time. The explanation for i_2 is that when v_L attempts to go to

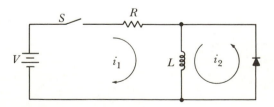

Figure 4.32   Free-wheeling diode circuit.

$-\infty$ as calculated, the diode will be forward-biased. It will turn on such that the current through L does not stop suddenly and no high voltage is generated.

EXERCISES

QUESTIONS

Q4.1 How does the breakdown region of a Zener diode differ from the breakdown region of an ordinary PN-junction diode?

Q4.2 How do the forward-conduction regions of Zener diodes and ordinary PN-junction diodes differ?

Q4.3 In what way is a backward diode a modification of a tunnel diode?

Q4.4 What are the desirable properties of tunnel diodes?

Q4.5 What is a good application for backward diodes?

Q4.6 How does a hot-carrier diode differ from a PN-junction diode

(a) In materials and/or construction?

(b) In terminal characteristics?

Q4.7 How do the characteristics of a snap diode differ from the characteristics of any other PN-junction diode?

Q4.8 What is the difference between a photodiode and a photocell?

PROBLEMS

P4.1 The component values for a voltage-regulator circuit are to be determined. The following conditions must be satisfied by the circuit.

$V_L = 12$ V, $R_L = 1$ kΩ, I_Z (min) $= 5$ mA

The circuit diagram is shown in Figure 4.33. Find

R_1 (max), P_{R1} (max), V_Z, P_Z (max)

P4.2 A proposed voltage-regulator circuit is shown in Figure 4.34. A cold-cathode voltage-regulating tube is available with specifications of $V_Z = 75$ V, and normal glow, 5 to 40 mA. The circuit is to maintain a constant 75 V and a constant 20 mA to a load resistor from a voltage source that varies between 100 and 135 V.

(a) Find R_1, P_{R1} (max), and V_Z.

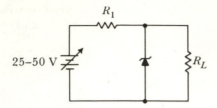

Figure 4.33   See Problem 4.1.

(b) Determine if the available regulator tube can fulfill the requirements of this application. If it cannot, completely specify a suitable Zener diode.

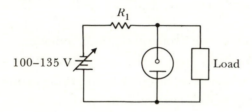

Figure 4.34   See Problem 4.2.

P4.3 The regulator circuit of Figure 4.35 is intended to supply a constant voltage to a varying load. The following parameters apply.

$V_{source} = 15$ V, $V_{load} = 8$ V,

$I_L = 10$ to 80 mA, I_Z (min) $= 3$ mA

Find values for the following quantities.

R_1 (max), P_{R1} (max), V_Z, P_Z (max)

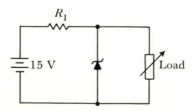

Figure 4.35   See Problem 4.3.

P4.4 Find all component values for the voltage-regulator circuit of Figure 4.36 to meet the following conditions.

$V_{source} = 35$ V, $V_{load} = 28$ V

$R_L = 50 \, \Omega$ to ∞, I_Z (min) $= 8$ mA

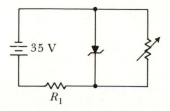

35 V

R_1

Figure 4.36   See Problem 4.4.

P4.5 The circuit of Figure 4.37 is to regulate in the presence of a varying source and a varying load. The following conditions apply:

$V_{source} = 35 - 50$ V, $V_L = 20$ V

$I_L = 0 - 25$ mA, I_Z (min) $= 5$ mA

Find the following values.

R_1, P_{R1} (max), V_Z, P_Z (max)

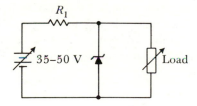

R_1

35–50 V   Load

Figure 4.37   See Problem 4.5.

P4.6 A voltage-regulator circuit is to be designed to meet the following conditions.

$V_{source} = 50 - 60$ V, $V_L = 32$ V

$R_L = 100 - 2000 \, \Omega$, I_Z (min) $= 10$ mA

    (a) Draw the appropriate circuit diagram.

    (b) Find R_1, P_{R1} (max), V_Z, P_Z (max).

P4.7 It is proposed to use the circuit of Figure 4.38 to smooth a partially filtered waveform from a rectifier circuit. The partially filtered voltage measures 135 V on a dc voltmeter. Within what range of values must the input-signal voltage be kept if the circuit is to properly regulate at 100 V? Conditions are as follows.

$V_L = 100$ V, $R_L = 1$ kΩ, I_Z (min) = 5 mA

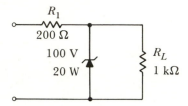

Figure 4.38   See Problem 4.7.

P4.8 Design a Zener-diode voltage-regulating circuit to regulate the voltage across a load at 12 V. The load current may vary from 20 to 120 mA. A 20-V dc source is available. Draw a circuit diagram and find all component values.

P4.9 An unregulated dc supply voltage varies over the range of 75 to 115 V. Design a Zener-diode regulating circuit to supply a regulated 50 V to a load that draws 10 mA. Draw the circuit diagram and find all needed component values.

P4.10 An unregulated dc voltage is available that varies between limits of 18 and 45 V. Design a Zener regulating circuit that will take the unregulated input and supply a regulated 10 V to a load resistance that varies between 100 Ω and infinity. Draw the circuit diagram and find all component values.

P4.11 An ac voltmeter circuit is to be built that will have a full-scale value of 135-V rms for sine-wave signals. A 1-mA 50-Ω meter movement is available. Assume an ideal diode.

(a) Draw the circuit diagram.

(b) Calculate all component values.

P4.12 Repeat Problem 4.11 if the full-scale reading is to be 250 V peak.

P4.13 A silicon *PN*-junction diode ($V_F = 0.6$ V) is to be used to protect a 1-mA 50-Ω meter movement. The circuit connection is shown in Figure 4.39. Calculate the value of R needed if the diode is to start bypassing current when the meter is carrying twice-rated current.

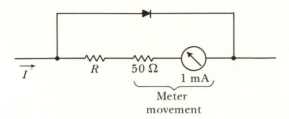

Figure 4.39   See Problem 4.13.

P4.14 Repeat Problem 4.13 if the protective diode is germanium with $V_F = 0.3$ V.

P4.15 A switch is used to disconnect a 600-V dc source from a load consisting of a 100-H inductance and a series 50-Ω resistance. A free-wheeling diode is to be used to protect the switch from arcing.

(a) Draw a diagram of the necessary circuit.

(b) Determine the PIV and forward-current ratings of the diode.

P4.16 Repeat Problem 4.15 if the load consists of a 500-H inductance in series with a 50-Ω resistance.

CHAPTER 5
BIPOLAR JUNCTION TRANSISTORS

The first transistors were developed at the Bell Telephone Laboratories by William Shockley, Walter Brattain, and John Bardeen. The new device was announced in 1948. These men received the Nobel physics award for their work in this endeavor. The word *transistor* is coined from the words *transfer resistor.*

The transistor has been vastly improved since its announcement in 1948. It has virtually replaced tubes in many, if not most, electronic applications. Only in certain specialized applications are vacuum tubes superior to present-day transistors.

5.1 CONSTRUCTION

The transistors we will be considering are called *bipolar junction transistors* (BJTs). Bipolar junction transistors have superseded an earlier type called the point-contact transistor.

Bipolar junction transistors have two *PN* junctions. They are three-terminal devices. The terminals are named the emitter, the base, and the collector. Figure 5.1 is a representation of the relationships between the junctions and the three terminals. Note that BJTs come in two distinct styles—the *PNP* and the *NPN* type. For each type there is a *PN* junction between the base and emitter leads and another *PN* junction between the base and collector leads. The word

151

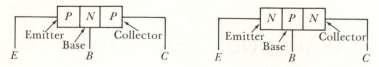

Figure 5.1   Representation of a bipolar junction transistor.

transfer was used in naming the transistor because the *I-V* characteristics of one junction are largely affected by the conditions prevailing
at, or transferred from, the other junction.

Bipolar junction transistors are usually made of the semiconducting materials germanium or silicon. There are specialized and complex processes used in the making of transistors. Somewhere in the
manufacturing processes the materials must be purified, then properly doped, and finally connected as a transistor. At this time it does
not seem appropriate to attempt a study of the many manufacturing
processes involved in making transistors.

The physical size of the semiconductor chip used in the making of
a transistor is rather interesting. The size of the chip will vary greatly
according to the intended application. BJTs meant for high-power
applications will have a relatively large semiconductor chip, that is, a
large cross-sectional area. A transistor that must operate at very high
frequencies will probably be made with a very small chip. BJT chips
normally range from about five-thousandths of an inch per side (of a
square) to about ¼ in. If you dismantle a transistor case to look at the
chip, you will find that the chip occupies a very small percentage of
the total transistor volume.

5.2   BIPOLAR JUNCTION TRANSISTOR *I-V* CHARACTERISTICS

A bipolar junction transistor has two *PN* junctions. A review of the
characteristics of *PN* junctions at this time should be helpful. Figure
5.2 shows *PN*-junction diodes in both the forward- and reverse-bias
directions. The charges available for conduction are holes in the *P*-
type material and electrons in the *N*-type material. When forward-
biased, as in Figure 5.2(a), the available charges are forced toward
the junction where the holes and electrons cancel. The *PN* junction,
then, can conduct quite easily when biased in the forward direction.
When reverse-biased, as in Figure 5.2(b), the available charges are
attracted away from the junction. The junction region has only
bound charges that are not available for conduction. Since there are
no charges crossing the junction, there is no current flow.

A BJT is a single crystalline structure containing two *PN* junctions. To get proper BJT action, one junction is forward-biased and the other is reverse-biased. An important point to remember is: For normal operation, a BJT is connected such that the emitter-base junction is forward-biased and the collector-base junction is reverse-biased.

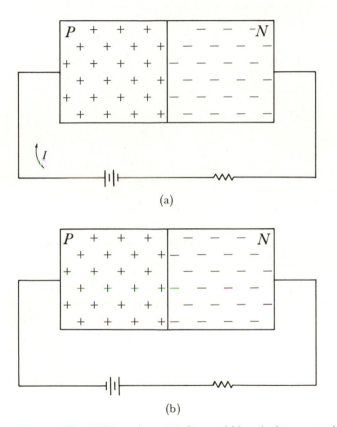

Figure 5.2   *PN* junctions: (a) forward-biased; (b) reverse-biased.

In order to understand BJT action, let us represent the current conduction of a *PN* junction somewhat differently than before. We will expand the new representation to thereby include the operation of a BJT. In Figure 5.3 we see that the *PN* junction is forward-biased and is conducting. This representation is different in that the current is shown as the clockwise movement of positive charges. Previously we have considered the charges to be holes in the *P*-type material and electrons in the *N*-type material. The two representations are equivalent with respect to charge, since a hole (absence of

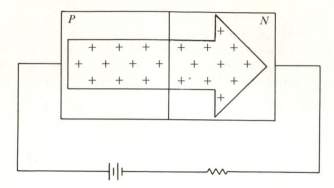

Figure 5.3   Pictorial representation of current flow in a *PN*-junction diode.

an electron) moving to the right is the same thing as an electron moving to the left.

Figure 5.4 is presented in order that we might have some way to visualize BJT action. The emitter-to-base junction is forward-biased and the collector-to-base junction is reverse biased. Since the emitter-base junction is forward-biased, we expect, and get, some current around the emitter-base loop. With the collector-base junction reverse-biased, we do not expect current flow in the collector-base loop. We do get some collector current, however, because the charges entering the base region, from emitter-base conduction, are attracted toward and across the collector-base junction by the higher collector voltage. (The reverse-biasing collector-base voltage is always of such a polarity as to attract these charges.) The emitter current is seen to "divide," or "branch," between the base and collector leads.

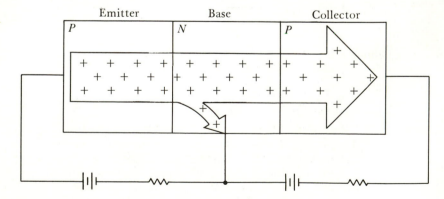

Figure 5.4   Pictorial representation of BJT action.

We can increase the base current of a BJT by increasing the forward bias on the emitter-base junction. When we do so, we find that the collector current increases proportionately. The ratio of collector current to base current is usually approximated as a constant value for any given transistor.

The overall operation of a BJT in a normal mode of operation is sometimes referred to as *transistor action* or, more properly, as *BJT action*. BJT action is a way of describing the division of emitter current into the base-current and collector-current components.

The BJT is useful as a control device because a small (base) current is able to control a much larger (collector) current. The ratio of the controlled (collector) current to the control (base) current may be as high as 1000. Values of 20 to 100 are very commonly available.

As mentioned before, BJTs can be made in two distinct types. These types are *NPN* and *PNP*. The type shown in Figure 5.4 is a *PNP* type. The operation of *NPN* BJTs is similar to the operation of *PNP* BJTs. Basically the types differ in that voltage and current polarities are opposite. In Figure 5.5 we have the pictorial representations and schematic symbols for both the *NPN* and *PNP* types.

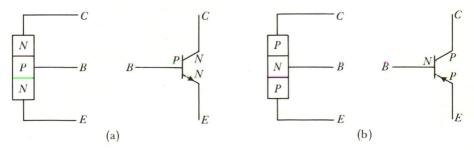

Figure 5.5 Bipolar-junction-transistor schematic symbols: (a) *NPN* type; (b) *PNP* type.

The current-flow differences between the two BJT types are as shown in Figure 5.6.

The three variables of greatest interest to us in the study of BJTs and BJT amplifiers are collector current, base current, and collector wrt emitter voltage. We are presented with the problem of displaying these three variables on a two-dimensional graph. The solution lies in the use of a family of curves.

A family of curves is constructed as follows. The base current is held at a constant value and data is taken to plot a graph of collector current I_C versus collector wrt emitter voltage V_{CE}. The base current I_B is then held at a different constant value while data is taken to plot

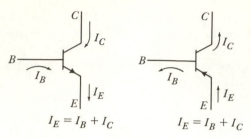

Figure 5.6   Current in BJTs; (a) *NPN* type; (b) *PNP* type.

another I_C-versus-V_{CE} plot on the same coordinate axis as before. By repeating this process a sufficient number of times, we arrive at a family of curves that does in fact display three variables on a two-dimensional graph.

The standard way to display the *I-V* characteristics of BJTs is to let the collector current I_C be the ordinate, the voltage V_{CE} be the abscissa, and the base current I_B be the so-called running or control variable. The *I-V* characteristics of a typical bipolar junction transistor might appear as in the family of curves shown in Figure 5.7.

According to standard mathematical convention, the *I-V* characteristics shown in Figure 5.7 must be for an *NPN* BJT. We know this to be a fact because V_{CE} is positive. The I_C and I_B quantities are also positive. If we were to follow normal mathematical conventions, we

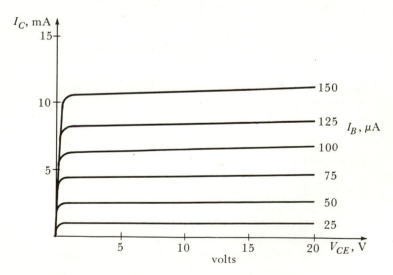

Figure 5.7   BJT *I-V* characteristics.

would find that the characteristics of *NPN* devices were drawn in the first quadrant and *PNP* devices drawn in the third quadrant of a standard cartesian coordinate system. For simplicity, however, we prefer to draw the characteristics for either type device in the first quadrant. A *PNP* device characteristic will plot in the first quadrant if instead of plotting I_C, I_B, and V_{CE}, we plot $-I_C$, $-I_B$, and $-V_{CE}$. In this book we will plot both *NPN* and *PNP* characteristics in the first quadrant. The variables I_C, I_B, and V_{CE} are to be interpreted as magnitudes only. Whether the device represented is *NPN* or *PNP* will have to be dealt with separately.

5.3   DYNAMIC PARAMETERS OF BIPOLAR JUNCTION TRANSISTORS

There are two dynamic parameters defined for transistors:

$$\alpha = \text{alpha} = \left. \frac{\Delta I_C}{\Delta I_E} \right|_{\Delta V_{CE}=0}$$

$$\beta = \text{beta} = \left. \frac{\Delta I_C}{\Delta I_B} \right|_{\Delta V_{CE}=0}$$

Note that in each case the quantities are ratios of currents. Thus the units of amperes cancel and the dynamic parameters are unitless (dimensionless) quantities.

The equations for alpha (α) and beta (β) may need some explanation. As in previous cases, the delta (Δ) symbol refers to a change in value. The vertical line is used to indicate that the portion to the left of the vertical line is to be evaluated under the conditions stated on the right side of the vertical line. As an example, we can state the alpha equation as follows: "Alpha is equal to a change in I_C divided by a change in I_E where the changes in I_C and I_E are evaluated while the change in V_{CE} is equal to zero." If the change in V_{CE} is equal to zero, then we know that V_{CE} must be a constant value. Thus the equation is sometimes written as

$$\alpha = \left. \frac{\Delta I_C}{\Delta I_E} \right|_{V_{CE}=k}$$

The dynamic parameter β can be evaluated numerically by a graphical construction on the characteristic curves of a particular BJT. In Figure 5.8 we have such a construction. In this case V_{CE} is

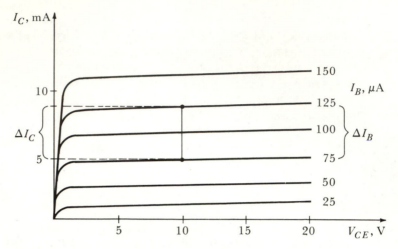

Figure 5.8   Graphical construction for the determination of beta (β).

constant at 10 V. Reading the numerical quantities from the graph, we have

$$\beta = \frac{\Delta I_C}{\Delta I_B}\bigg|_{V_{CE}} = 10 \text{ V} = \frac{(8.5 - 4.5) \text{ mA}}{(125 - 75) \text{ }\mu\text{A}}$$

$$= 80$$

The BJT is a good current-controlling device. The BJT whose characteristics are shown in Figure 5.8 has a controlled current 80 times as large as the controlling current. In its most common connection this particular transistor has a current gain of 80.

The parameter α cannot be determined graphically from the characteristic curves, as shown in Figure 5.8. We prefer to evaluate α by noting the relationships between the defining equations for α and β.

$$\beta = \frac{\Delta I_C}{\Delta I_B}$$

$$\alpha = \frac{\Delta I_C}{\Delta I_E} \quad \text{or} \quad \Delta I_C = \alpha \, \Delta I_E$$

Substituting for known quantities in the β equation, we get

$$\beta = \frac{\alpha \, \Delta I_E}{\Delta I_B} = \frac{\alpha \, \Delta I_E}{\Delta (I_E - I_C)} = \frac{\alpha \, \Delta I_E}{\Delta I_E - \Delta I_C}$$

$$= \frac{\alpha \, \Delta I_E}{\Delta I_E - \alpha \, \Delta I_E} = \frac{\alpha \, \Delta I_E}{\Delta I_E (1 - \alpha)} = \frac{\alpha}{1 - \alpha}$$

By rearranging terms we can solve for α:

$$\alpha = \frac{\beta}{1 + \beta}$$

In this particular example,

$$\alpha = \frac{80}{1 + 80} = 0.987654320987\ldots$$

(Would you believe that this decimal quantity for α is a coincidence?)

BJT action involves the division of the emitter current into the base and collector currents. Since I_C is always smaller than I_E, bipolar junction transistors must have an α of less than unity.

$$\alpha = \frac{\Delta I_C}{\Delta I_E}\bigg|_{V_{CE}=k} < 1$$

5.4   BASIC-AMPLIFIER CONNECTION AND THE LOAD LINE

A bipolar junction transistor can be used with other components in such a way as to form an amplifying circuit. In Figure 5.9, we see examples of BJT amplifier circuits, each of which is the same basic configuration with slightly different base-current sources. The BJTs are *NPN* type. Notice that the output signal v_{OUT} is taken between

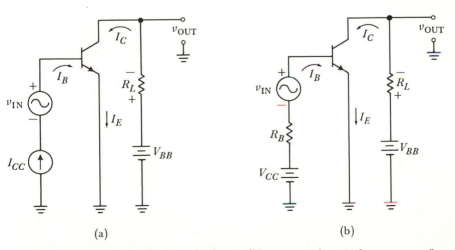

(a)   (b)

Figure 5.9   *NPN* BJT in basic amplifier connection: (a) base current from current source; (b) base current from voltage source and resistance.

collector and ground. The input signal v_{IN} is applied essentially between base and ground. The emitter is connected directly to ground.

The circuits of Figure 5.9 represent the BJT version of what we choose to call the *basic amplifier connection.* Although this circuit has another name (common emitter), we prefer the name basic amplifier connection at this time. The basic amplifier connection is without doubt the most widely used of all possible amplifier-circuit configurations. We want to study this connection thoroughly and have plenty of time for the knowledge to sink in before we acknowledge that any other amplifier-circuit configurations exist. Since the basic amplifier connection is used in perhaps 90% of all amplifier applications, we will spend perhaps 90% of our time and effort in the study of this common circuit.

The basic amplifying circuit can also be built using *PNP* BJTs. The basic difference between *NPN* and *PNP* devices is the change of polarity of all voltages and currents. In Figure 5.10 we have *PNP* BJTs connected in the basic-amplifier configuration. Note the polarities of all circuit voltages and currents. Emitter current ($I_E = I_C + I_B$) is in the direction of the arrow on the transistor symbol. From Figures 5.9 and 5.10 we see that the output voltage v_{OUT} is equal to v_{CE} for this connection.

We are now in a position to write some loop equations that describe the basic amplifier circuit. However, before we write the equations we want to comment on the quantities to be used.

The symbols V_{CC}, and V_{RL}, and I_C convey no information about the

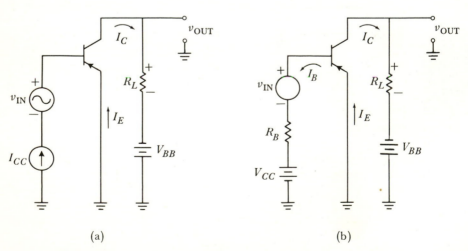

Figure 5.10 *PNP* BJT in basic amplifier connection: (a) base current from current source; (b) base current from voltage source and resistor.

polarity of the quantity represented. If the symbols do not convey polarity information, how then can appropriate polarities be used when writing equations? We can answer that question by referring to Figure 5.9. The reference direction of V_{CC} is defined by the way the battery symbol is drawn. The reference direction of current I_C is defined by the arrow representing I_C. The voltage across R_L is of course $I_C R_L$ and its reference direction must be consistent with the defined direction of I_C.

Since students often get confused by the items mentioned in the preceding paragraph, let us practice writing equations for the circuit of Figure 5.9(b) and 5.10(b). To emphasize the points just made, we will enclose the V_{CC} and I_C symbols in absolute-value bars. Starting at the ground symbol off the emitter in Figure 5.9(b) and going clockwise to write the KVL equation, we obtain

$$+V_{CE} + |I_C|R_L - |V_{BB}| = 0$$

Starting at the same point and going counterclockwise, we obtain

$$+|V_{BB}| - |I_C|R_L + V_{EC} = 0$$

Both these equations apply to the same circuit and of course are identical. At first glance, however, it appears that the quantities I_C and V_{BB} have different polarities depending on the path taken when writing the equation.

If we now write the clockwise and then counterclockwise KVL equations for the circuit of Figure 5.10 in the same manner as we did for Figure 5.9, we obtain

$$+V_{CE} - |I_C|R_L + |V_{BB}| = 0$$

and

$$-|V_{BB}| + |I_C|R_L + V_{EC} = 0$$

Another common source of confusion and error concerns double-subscripted variables such as V_{CE} in the equations above. V_{CE} appears in the clockwise-obtained equations for both the *NPN* and *PNP* case. In each case it is preceded by a plus sign. Actually V_{CE} would be a positive voltage for the *NPN* case and a negative voltage for the *PNP* case. The point to be made is that a variable such as V_{CE} may represent any quantity (positive, negative, or zero) regardless of the sign preceding it in an equation.

The items mentioned in the last several paragraphs really should not need inclusion here. Experience has shown, however, that there is a considerable amount of confusion about writing loop equations at this point. It is suggested that every student stop now and write the KVL equations for both the *NPN* and *PNP* cases and for both clockwise and counterclockwise directions. If your equations agree with those shown above, then in fact this discussion was unnecessary. If your equations do not agree, however, give as much thought or get as much help as necessary to clear up the situation.

Now we are ready to continue with the writing of KVL equations for the circuits of Figures 5.9 and 5.10. We will drop the use of absolute-value bars and proceed with confidence. For the *NPN* case (Figure 5.9) we obtain

$$V_{CE} + V_{RL} - V_{BB} = 0$$
$$V_{CE} = V_{BB} - V_{RL}$$

For the *PNP* transistor case (Figure 5.10) we find that

$$V_{CE} - V_{RL} + V_{BB} = 0$$
$$V_{CE} = -V_{BB} + V_{RL}$$
$$-V_{CE} = V_{BB} - V_{RL}$$

Note that the polarity associated with V_{CE} differs depending on whether the circuit contains an *NPN* BJT or a *PNP* BJT.

In Section 5.2 when we were discussing *I-V* characteristics of BJTs, we decided to draw the characteristics in the first quadrant regardless of transistor polarity. In order to be consistent with the earlier decision, we will now assume that the equation for V_{CE} is meant to imply magnitude only. Thus we write

$$V_{CE} = V_{BB} - V_{RL}$$

In the case of an *NPN* BJT, the equation is correct as written. In the case of the *PNP* BJT, however, we must realize that the numerical value of the V_{CE} quantity will have a negative sign associated with it.

No doubt the reader thinks that we are going through a lot of unnecessary work. This extra work, however, allows us to draw the characteristics and the forthcoming load line in the same manner regardless of the transistor polarity. It is a convenience.

The equation we have developed will lead to something called a *load-line equation*. By substituting for V_{RL} we obtain

$$V_{CE} = V_{BB} - V_{RL} = V_{BB} - I_C R_L$$

This equation relating V_{CE} and I_C is called the load-line equation. In any given BJT amplifier circuit, we would have specific constant values for V_{BB} and for R_L. The load-line equation is thus a linear equation in the two variables V_{CE} and I_C. It is normally plotted superimposed on a family of curves for the particular BJT used in the circuit. Such a plot is, of course, called the load line of the circuit. Figure 5.11 shows a BJT amplifier circuit with its corresponding characteristic curves and load line.

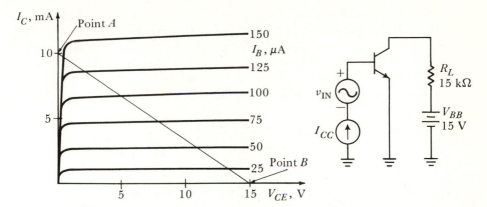

Figure 5.11   BJT amplifier and load line.

We can see by inspection that the equation describing the load line is linear. The load line is therefore known to be a straight line, and any two points on the line will define the line. The endpoints of the load line are easy to find. When $I_C = 0$ we have

$$V_{CE} = V_{BB} - I_C R_L = V_{BB} - 0$$

$$= V_{BB}$$

Thus one endpoint is described as

$$I_C = 0, \qquad V_{CE} = V_{BB}$$

When $V_{CE} = 0$ we have

$$V_{CE} = 0 = V_{BB} - I_C R_L$$

$$V_{BB} = I_C R_L$$

$$I_C = \frac{V_{BB}}{R_L}$$

The other endpoint is $V_{CE} = 0$, $I_C = V_{BB}/R_L$.

For the particular example in Figure 5.11, the endpoints are as follows:

Point A:

$$V_{CE} = 0, \qquad I_C = \frac{15 \text{ V}}{1.5 \text{ k}\Omega} = 10 \text{ mA}$$

Point B:

$$I_C = 0, \qquad V_{CE} = 15 \text{ V}$$

The load-line equation gives the mathematical relationship between V_{CE} and I_C. The load line, constructed on the characteristic curves, gives the same information graphically. The operating point of the circuit at any and all instants of time must be somewhere on the load line. Much information about the operation of the circuit can be obtained graphically from the load-line construction.

5.5   THE QUIESCENT POINT

Quiescent point refers to the steady-state operating conditions existing throughout the circuit. The quiescent point (Q point or quiet point) describes the operating point when no signals are present. Since the Q point is a circuit operating point, it must appear somewhere on the load line. When signals are present, we find that the signal levels cause the instantaneous operating point to deviate from and around the dc operating point, which is the Q point. The operating point at any instant of time is the superposition of the Q point and the instantaneous value of the appropriate signal level.

The Q point of a BJT amplifier circuit is to be determined by the circuit designer. To choose an appropriate Q point intelligently, the designer must have one or more criteria upon which to base a choice. Three possible criteria that would affect the location of the optimum Q point are as follows:

1. Magnitude of expected signals.

2. dc power-dissipation considerations.

3. Linearity of the circuit.

Let us consider the implications of each of these criteria separately.

Suppose that we are considering an appropriate choice for the Q point of the circuit of Figure 5.12. If we want the circuit to be able to handle output-signal levels of at least 15 V peak to peak, for example, then we know that the supply voltage V_{BB} must be somewhat greater than 15 V. To be safe, let us choose a supply voltage of 20 V. Based on the 20 V supply voltage, there are three separate load lines drawn. (An infinite number of load lines could be drawn.) Each of these load lines represents a circuit capable of handling output-signal (ΔV_{CE}) levels of 15 V peak to peak. We will arbitrarily choose the load line that goes through the point $I_C = 10$ mA, $V_{CE} = 0$ and labeled as load line 2. For this to be so, we must make $R_L = 2$ kΩ.

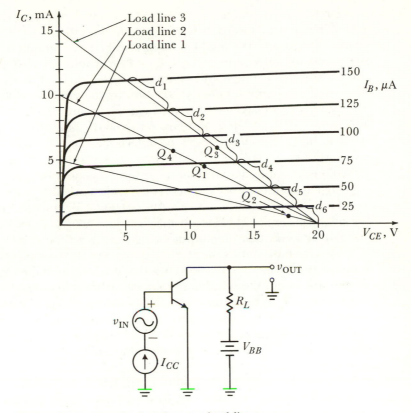

Figure 5.12 Q point locations on load line.

Since the output voltage (that is, V_{CE}) must be able to swing 7.5 V on either side of the Q point (total of 15 V peak to peak), the Q point must be at least 7.5 V from either useful end limit on the load line. One end limit is where $V_{CE} = 20$ V. The other end limit appears to be near the point where V_{CE} is about 2 V. For V_{CE} less than about 2 V, the transistor becomes quite nonlinear, as indicated by the different shape of curves for V_{CE} values less than about 2 V. If we choose midway between the end limits as the Q point, the amplifier circuit will be able to take equal swings in each direction before distortion occurs. Therefore, let us choose the Q point as the point on the load line where $V_{CE} = 11$ V. This point is shown in Figure 5.12 as point Q_1. This circuit can handle an output signal of 18 V peak to peak, that is, a swing of 9 V in each direction.

A circuit that is optimized in terms of minimum dc power dissipation will in general have a different Q point than the Q_1 just decided upon. The power dissipated by the BJT will be a minimum (zero) at the points on the load line where $I_C = 0$ or where $V_{CE} = 0$, that is, the ends of the load line. Minimum power is dissipated by R_L when $I_C = 0$. For minimum power dissipation by the entire circuit, then, we will want to place the Q point near the point on the load line where $I_C = 0$, $V_{CE} = 20$ V. Just how far from the point $I_C = 0$, $V_{CE} = 20$ that the Q point must be will be determined by the amplitude of the output-signal voltage that this circuit must handle. To present a ridiculous example, assume that the output-voltage swing will never exceed 1 mV peak to peak. Theoretically, then, we can place the Q point on the load line where $V_{CE} = (20 - 0.0005)$ V. More practically, we might place the Q point on the load line at $V_{CE} = 18$ V, even though we know that the output-voltage swing will be small.

Unless we have some other restraints, we can save some power by building the circuit to have a load line that goes through the $V_{CE} = 0$, $I_C = 5$ mA point. We will choose to do so (labeled load line 1). This will require a load resistance of 4 kΩ. We will in this case choose a Q point at $V_{CE} = 18$ V on load line 1. This point is labeled Q_2 in Figure 5.12. You must realize that we could also have chosen a larger R_L (I_C intersection at 1 or 2 mA) or a small value for V_{BB}, since the signal levels are small.

Suppose that we want to optimize the circuit of Figure 5.12 in terms of linearity. The family of base currents shown differ by the constant magnitude of 25 μA. If the circuit is linear, the base-current lines will cross the load line at equidistant intervals. These intervals are labeled d_1 through d_6 on the figure. Although d_1 and d_6 are definitely different size intervals, it appears that intervals d_2, d_3, and d_4 are approximately equal. The points Q_1 and Q_3 seem to be quite satisfactory in terms of obtaining a circuit with good fidelity (linearity). Drawing another set of intersection intervals will show that Q_1 is not quite in the middle of the most nearly linear region. Therefore, for fidelity considerations, let us assume that Q_3 is optimal.

There are possible reasons to choose a Q point on some other load line with an intersection of greater than the 15 mA shown for load line 3. The necessity of a low output impedance would dictate such a preference. A low output impedance means a low Thévenin-equivalent resistance. Controlling the Thévenin resistance by choosing the correct load line may allow us to increase the signal transferred to any following stage. An example relating to impedances is included in Chapter 8.

5.6 BJT BIASING METHODS

We have talked about BJTs connected in amplifier circuits. We have also considered where the Q point should be located for a particular amplifier circuit to meet specified requirements. The problem remaining is to find circuits or techniques that will require the circuit to operate at the particular point we desire. These circuits or techniques are the bias circuits or bias techniques. When we speak of the BJT bias, we are referring to the dc or nonsignal conditions existing throughout the circuit.

BJT amplifier circuits have been discussed several times already in this chapter. In some of these cases, the base current was specified as coming from a current source in the base circuit. The current source is fine from a theoretical point of view. In practical terms, voltage sources are used much more often than current sources in the field of electronics. For example, a voltage source is needed in the collector-emitter circuit of the BJT basic amplifier circuit. It seems appropriate, therefore, to discuss only biasing techniques that somehow make use of constant-voltage sources.

The purpose of any bias circuit is to cause a BJT circuit to operate at some chosen and specified Q point. In order to simplify the numbers involved, let us talk about the same BJT and bias it at the same point as we discuss the various biasing techniques and circuits. In Figure 5.13 we see the characteristics of the BJT, the chosen load line and the chosen Q point. Note that by inspection of the load line

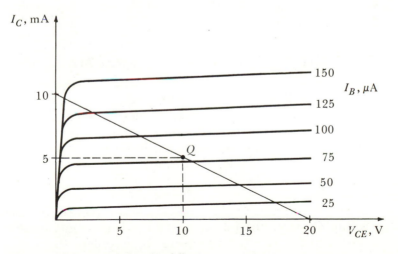

Figure 5.13 Q point on load line.

we can determine that $V_{BB} = 20$ V and $R_L = 2k$. We will assume that at the Q point

$$V_{CE} = 10 \text{ V}$$

$$I_C = 5 \text{ mA}$$

$$I_B = 80 \text{ }\mu\text{A}$$

FIXED BIAS

The first bias-circuit connection we will discuss is the two-battery bias circuit. It is a form of fixed bias. This connection is shown in Figure 5.14. Separate batteries are used for the collector supply and the base supply. The numerical values of V_{BB} and R_B must be chosen so that $I_B = 80$ μA. We probably want to use a standard value for the base battery and then choose R_B to supply the correct base current.

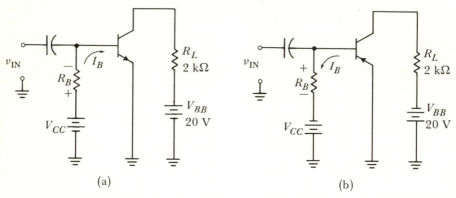

(a)   (b)

Figure 5.14   Two battery bias methods: (a) *NPN* BJT; (b) *PNP* BJT.

The calculations involved in choosing R_B must take into account the base-emitter voltage of the BJT. The base-emitter junction is, of course, forward-biased. If we assume that the junction voltage is zero (that is, an ideal junction) then of course the base current is

$$I_B = \frac{V_{CC}}{R_B}$$

If we are interested in the accuracy of the Q-point placement, we should consider the turn-on voltage of the emitter-base junction. The turn-on voltage of the junction is dependent on the material of which the BJT is made. Normally we assume that the turn-on voltage of a silicon *PN* junction is 0.6 V and that of a germanium *PN* junction is 0.3 V. Writing the Kirchhoff-voltage-law equation around the loop including the emitter-base junction, we obtain

$$V_{CC} - V_{RB} + V_{EB} = 0 \qquad \text{for } NPN \text{ BJT}$$

$$-V_{CC} + V_{RB} + V_{EB} = 0 \qquad \text{for } PNP \text{ BJT}$$

If we are considering silicon BJTs, we have

$$V_{CC} - V_{RB} - 0.6 = 0 \qquad \text{for } NPN \text{ BJT}$$

$$-V_{CC} + V_{RB} + 0.6 = 0 \qquad \text{for } PNP \text{ BJT}$$

These equations are the same and can be written as $|V_{RB}| = |V_{CC}| - 0.6$. Assume now that we decide to use a 6.3 V battery as the base-supply source V_{CC}. We can solve for I_{RB}, which is the same as I_B.

$$I_{RB} = I_B = \frac{V_{RB}}{R_B} = \frac{(6.3 - 0.6) \text{ V}}{R_B}$$

We have already determined that I_B should be 80 μA for our example problem. Thus we have

$$I_B = 80 \ \mu\text{A} = \frac{(6.3 - 0.6)}{R_B}$$

or

$$R_B = \frac{(6.3 - 0.6)}{80 \ \mu\text{A}} = \frac{5.7 \text{ V}}{80 \ \mu\text{A}} = 71.2 \text{ k}\Omega$$

If we were to choose a germanium BJT, the calculated value of R_B would be somewhat different. The reason for the difference, of course, is the difference in the emitter-base turn-on voltage. In the case of a germanium BJT, we have

$$V_{RB} = V_{CC} - 0.3$$

Using the same battery voltage of 6.3 V, we obtain

$$I_{RB} = \frac{(6.3 - 0.3) \text{ V}}{R_B}$$

Since we need 80 μA of base current, we have

$$I_{RB} = I_B = 80 \ \mu\text{A} = \frac{6.0 \text{ V}}{R_B}$$

$$R_B = \frac{6.0 \text{ V}}{80 \ \mu\text{A}} = 75 \text{ k}\Omega$$

The two-battery method is of course capable of biasing a BJT. Since we were able to choose the base-battery voltage, it would seem that we could use the same battery voltage for the base supply as for the collector supply. Not only can we use the same voltage magnitude for the two supplies, we can use the same battery. Thus we have the one-battery method of biasing a BJT. In Figure 5.15 we see the one-battery method for each polarity of BJT. Using the same BJT and

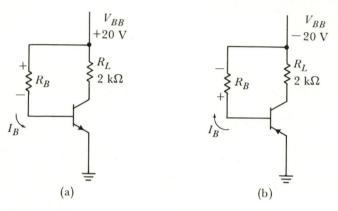

(a)   (b)

Figure 5.15   One-battery biasing method: (a) *NPN* BJT; (b) *PNP* BJT.

the same Q point shown in Figure 5.13, we can solve for the needed value of R_B. The equations for the voltage across R_B will be the same for the one-battery method as for the two-battery method. To show that this is so, we write the equation for the loop that includes the emitter-base junction in Figure 5.15. We obtain

$$+20 - V_{RB} + V_{EB} = 0 \qquad \text{for } NPN \text{ BJT}$$

and

$$-20 + V_{RB} + V_{EB} = 0 \qquad \text{for } PNP \text{ BJT}$$

Substituting the correct value of V_{EB} (-0.6 V for *NPN* silicon and $+0.6$ V for *PNP* silicon), we obtain

$$V_{RB} = 20 - 0.6 \qquad \text{for } NPN \text{ BJT}$$

and

$$V_{RB} = 20 - 0.6 \qquad \text{for } PNP \text{ BJT}$$

We can use the results just obtained to find a numerical value for R_B such that the base current $I_B = 80 \ \mu A$. Thus

$$I_{RB} = I_B = 80 \ \mu A = \frac{V_{RB}}{R_B} = \frac{20 - 0.6}{R_B}$$

$$R_B = \frac{(20 - 0.6)\ \text{V}}{80\ \mu\text{A}} = 242\ \text{k}\Omega$$

For the germanium BJT, we would obtain

$$R_B = \frac{(20 - 0.3)\ \text{V}}{80\ \mu\text{A}} = \frac{19.7}{80\ \mu\text{A}} = 246\ \text{k}\Omega$$

COLLECTOR BIAS

The biasing circuits we have analyzed thus far are capable of biasing the BJT exactly where we want it if the BJT is accurately represented by the characteristic curves we are using. However, when we obtain a set of characteristic curves from a BJT manufacturer, we are actually getting typical characteristics for BJTs of a given type rather than those for one specific BJT. The variation of the characteristics from the typical values may be rather large. In our example problem, we needed a base current of 80 μA for the specified Q point. If the data we have is given as typical for a given type, we might find some BJTs of the same type that require 160 μA and some that require 50 μA. Hence, the characteristics do vary over a considerable range. In addition to parameter variations from unit to unit in a batch of BJTs, the parameters change with temperature. If a BJT is biased properly at room temperature, it may not be biased properly at higher (or lower) temperatures. The biasing circuits we have discussed cannot compensate for any parameter variations.

There are biasing techniques that make the circuit somewhat self-adjusting if and when any parameter variations take place. Such circuits have what is called negative feedback. One such bias circuit is shown in Figure 5.16. We call this the collector bias method. This

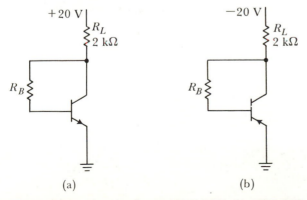

Figure 5.16   Bias circuit: (a) *NPN* BJT; (b) *PNP* BJT.

circuit will have much better bias stability than either of the bias circuits mentioned previously. If $V_{CE} = 10$ (or -10 for the *PNP* case), then the top of resistor R_B is at 10 V. The bottom of R_B is at the turn-on voltage of the emitter-base junction. Thus we know the voltage across R_B and can solve for R_B for whatever base current we want. $I_{RB} = I_B$.

We must understand the self-adjusting feature of this circuit before calculating any numerical values for our example bias point. Suppose that $V_{CE} = 10$ and R_B is calculated to supply the necessary 80 μA of base current. If for any reason β increased, the collector current would increase for the same 80 μA of base current. However, when the collector current increased, V_{CE} decreases owing to the increased voltage drop across R_L. The decreased V_{CE} causes a decrease in the voltage drop across R_B. The current in R_B must decrease also. Thus we see that as collector current increases due to increased β, the base current decreases, which helps keep the collector current near its former, and proper, operating-point value. A similar correcting action will occur for other variations of parameters.

Now let us calculate the necessary value for R_B so that the circuit will be biased at the Q point where

$V_{CE} = 10$ V

$I_C = 5$ mA

$I_B = 80$ μA

By inspection of the circuits shown in Figure 5.16, we have

$V_{RB} = 10 - 0.6$ for silicon BJT

$V_{RB} = 10 - 0.3$ for germanium BJT

These equations are true for both *PNP* and *NPN* BJTs. Solving for base current, we obtain

$I_B = I_{RB} = 80$ μA

$I_{RB} = 80$ μA $= \dfrac{V_{RB}}{R_B}$

$R_B = \dfrac{9.4 \text{ V}}{80 \text{ } \mu A} = 117.5$ kΩ for silicon BJT

$R_B = \dfrac{9.7 \text{ V}}{80 \text{ } \mu A} = 121.25$ kΩ for germanium BJT

Now that we have calculated circuit component values, we must point out that what we have done is not exactly right. We cannot simultaneously have $V_{CE} = 10$ and $I_C = 5$ mA without changing V_{BB} or R_L. The reason for this state of affairs is that in this circuit, I_{RL} is the sum of I_C and I_B. If we in fact insist that V_{CE} be exactly 10 V, then $I_{RL} = 5$ mA $= I_C + I_B$ or $I_C = 5$ mA $- I_B$. Since the base current I_B will be near but not exactly 80 μA, we find that $I_C \approx 5$ mA $- 0.08$ mA $= 4.92$ mA. Similarly, we will find that if we insist on having I_C exactly 5 mA, then V_{CE} will not be exactly 10 V. The errors introduced into the problem are very small. The method of solution seems justified in view of the very small errors involved.

UNIVERSAL-BIAS CIRCUIT

The last of the biasing circuits we will consider, the universal-bias circuit, also has the self-adjusting feature. The circuit connection is shown in Figure 5.17. Note that R_L has been decreased by 500 Ω and a 500-Ω resistor has been added in the emitter circuit. The added emitter resistor is needed. The collector resistor was changed in value so that the total collector-loop resistance remains 2 kΩ and the load line is not changed from the previous examples. If the emitter resistor had been added without reducing R_L, a new load line would have been needed. The new load-line intersection would have been, in this case, 20 V divided by 2.5 kΩ, which is 8 mA. Obviously with a new load line, a new Q point would have to be chosen.

We will now calculate values for R_1 and R_2 to make the circuit of Figure 5.17 operate at our chosen Q point. Since the total collector-loop resistance is still 2 kΩ, we design for our original Q-point choice; that is,

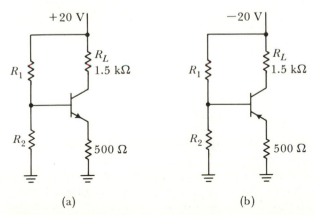

Figure 5.17 Bias circuit: (a) *NPN* BJT; (b) *PNP* BJT.

$$V_{CE} = 10 \text{ V}$$

$$I_C = 5 \text{ mA}$$

$$I_B = 80 \text{ } \mu\text{A}$$

By Ohm's law we can find the emitter voltage wrt ground. This voltage is the same as the voltage across the 500-Ω emitter resistor.

$$V_{500} = I_{500}500$$

$$= I_E 500 = (I_C + I_B)500$$

$$= [(5 + 0.08) \text{ mA}](500)$$

$$= 2.54 \text{ V}$$

(In some cases we might assume, for ease, that $I_C = I_E$. This assumption is nearly true.) The emitter voltage (wrt ground) is, then, $V_E = 2.54$ V. The voltage is strictly true only for the *NPN* case. For the *PNP* case, the polarity sign is the opposite one.

This problem will be solved in terms of the *NPN* BJT. The *PNP* solution will yield exactly the same numerical results with opposite polarity sign on all voltages and currents. Since we now know the emitter voltage (wrt ground), we can also find the base voltage (wrt ground).

$$V_B = V_E + V_{BE} \text{ turn on}$$

$$V_B = 2.54 + 0.6 = 3.14 \text{ V} \qquad \text{for silicon BJT}$$

$$V_B = 2.54 + .3 = 2.84 \text{ V} \qquad \text{for germanium BJT}$$

We are now in a position to calculate the magnitude of the voltage drops across R_1 and R_2. Resistor R_2 is connected from base to ground. Thus,

$$V_{R2} = V_B = 3.14 \text{ V} \qquad \text{for silicon BJT}$$

$$V_{R2} = V_B = 2.84 \text{ V} \qquad \text{for germanium BJT}$$

Resistor R_1 is connected between a 20-V source and base. Thus,

$$V_{R1} = 20 - 3.14 = 16.86 \text{ V} \qquad \text{for silicon BJT}$$

$$V_{R1} = 20 - 2.84 = 17.16 \text{ V} \qquad \text{for germanium BJT}$$

We now know the magnitude of the voltage drop across each of the bias resistors. Before we can calculate numerical values of R_1 and R_2, we must decide how much current each of these resistors is to carry. In Figure 5.18, we have the bias circuits redrawn with the currents in the bias resistors defined. Note that in each case $I_1 = I_B + I_2$. The only one of these currents specified by our choice of Q point is the base current I_B. Thus $I_1 = 80 \text{ } \mu\text{A} + I_2$. This circuit will

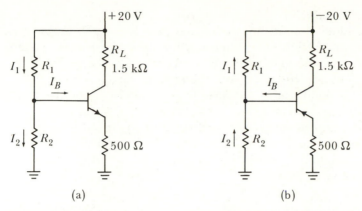

Figure 5.18   Bias-circuit currents: (a) *NPN* BJT; (b) *PNP* BJT.

work in a quite satisfactory manner if we let I_2 vary anywhere from zero to several times as large as I_B. For the case where the BJT parameters may change, it has been found that I_2 should be greater than zero. From past experience in building BJT-biasing circuits of this type, we have developed a rule of thumb for determining the value of I_2.

RULE   For good bias stability with a bias network as shown in Figure 5.18, the current I_2 should be two to five times as large as I_B. The higher value of I_2 has the better bias stability.

For our example let us choose I_2 to be twice as large as I_B. We can now find numerical values for I_1 and I_2.

$$I_2 = 2I_B = 160 \ \mu A$$

$$I_1 = I_2 + I_B = (160 + 80) \ \mu A = 240 \ \mu A$$

Now we know both the voltage across and the current in each of the bias resistors. We can calculate the numerical values as follows:

$$R_2 = \frac{V_{R2}}{I_{R2}} = \frac{3.145}{160 \ \mu A} = 19.7 \ k\Omega \qquad \text{for silicon BJT}$$

$$R_2 = \frac{V_{R2}}{I_{R2}} = \frac{2.84}{160 \ \mu A} = 17.75 \ k\Omega \qquad \text{for germanium BJT}$$

$$R_1 = \frac{V_{R1}}{I_{R1}} = \frac{16.86}{240 \ \mu A} = 70.2 \ k\Omega \qquad \text{for silicon BJT}$$

$$R_1 = \frac{V_{R1}}{I_{R1}} = \frac{17.16}{240 \ \mu A} = 71.5 \ k\Omega \qquad \text{for germanium BJT}$$

A word of caution is in order regarding circuits that have an emitter resistor. It is mentioned here because the last bias circuit we discussed has an emitter resistor. With an emitter resistor in the circuit, the output-voltage swing is not the same as the V_{CE} swing along the load line. Since the output voltage is taken from collector to ground, the output is really the sum of V_{CE} and the voltage across the emitter resistor. In terms of signal, V_{CE} and the emitter-resistor voltage change in opposite directions, so the output voltage is less for a circuit with an emitter resistor. If we can find the peak-to-peak swing in V_{CE} from the load line, we can find the output voltage as follows:

$$V_{\text{OUT}} = V_{CE}\left(1 - \frac{R_E}{R_E + R_L}\right)$$

Of the four bias circuits we have discussed, the last two are quite advantageous. The advantage is of course that once the circuit is designed, the Q point will be relatively stable even if we use a BJT that has a different β than the typical unit or if β changes when the circuit is operated throughout various temperature ranges. A disadvantage of these last two circuits is that the self-adjusting, or negative-feedback, feature also causes the circuits to have a somewhat lower voltage gain as compared with circuits without the self-adjusting feature. BJT circuits in high-quality equipment will nearly always use bias circuits with the self-adjusting feature. In many cases the bias circuits you see in commercially manufactured equipment will be just like the ones we have discussed. In other cases the bias circuits will be similar but not exactly like the circuits we have studied. The last-studied circuit is believed to be the most commonly used in high-quality equipment.

5.7 BASIC-AMPLIFIER CONNECTION: EQUATIONS AND WAVEFORMS

We have already talked about the BJT as connected in a basic-amplifier configuration. We have considered the optimum Q point for a BJT amplifier to meet certain specified criteria. We have studied several ways of biasing the BJT so that it does, in fact, operate at the Q point we have chosen. Now that we have taken care of these preliminaries, we want to look at the various voltage and current signals throughout the circuit. We will want to examine these signals on the basis of the mathematical equations involved and also in terms of the graphical plots of the waveforms.

Let us first consider an *NPN* BJT in a basic-amplifier connection with fixed bias. This circuit is shown in Figure 5.19 along with the associated characteristic curves and the load line. Suppose that we have a sinusoidal input voltage that results in a sinusoidal base current of 25 μA peak. The operating point of the circuit would vary between point A and point B on the load line, as shown in Figure 5.19. Mathematically speaking, we can say that $i_B = (80 + 25 \sin \omega t)$ μA. By reading numerical values as accurately as possible from the load line, we can say that $i_C = (5 + 2 \sin \omega t)$ mA and $v_{CE} = (10 - 4 \sin \omega t)$ V. The reason for the minus sign in the equation for v_{CE} can be seen by examining the load line. As the input signal changes from zero in the positive direction, the instantaneous operating point goes from Q toward point A on the load line. Note that as we move along the

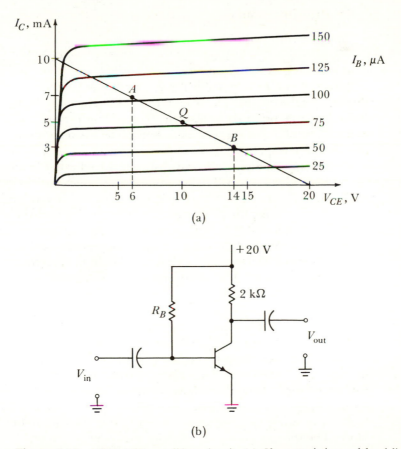

(a)

(b)

Figure 5.19 *NPN* BJT amplifier circuit. (a) Characteristics and load line. (b) Circuit connection.

load line from Q toward point A, the base current is increasing, the collector current is increasing, and v_{CE} is decreasing. The decrease in v_{CE} is shown by the minus sign in the equation.

As a kind of check on the v_{CE} equation in terms of numerical values and the unexpected minus sign, let us find v_{CE} by another method. By inspection of the circuit shown in Figure 5.19(b) we can say that

$$v_{CE} = 20 - v_{2\ k\Omega}$$

$$= 20 - i_C \times 2\ k\Omega$$

Substituting for i_C as taken from the load line, we have

$$v_{CE} = 20 - (5 + 2 \sin \omega t)\ \text{mA}\ (2\ k\Omega)$$

$$= 20 - 10 - 4 \sin \omega t$$

$$= (10 - 4 \sin \omega t)\ \text{V}$$

The result is gratifying, isn't it?

We are now able to graphically display the waveforms of interest throughout the circuit. The waveforms are as shown in Figure 5.20. Note the variation of dc levels when comparing v_{CE} and v_{OUT} waveforms. Note also that there is a phase reversal between input voltage and output voltage.

We also want to look at the waveforms involved with the same type circuit but with a *PNP* BJT. The circuit and characteristic curves are as shown in Figure 5.21. Note that the characteristic curves and load line look exactly like the *NPN* case. The reason they look the same is that we have plotted $-I_B$, $-I_C$, and $-V_{CE}$. Making the assumption that the input sinusoidal voltage is large enough to cause a sinusoidal signal current of 25 μA peak, we find that the circuit waveforms are as shown in Figure 5.22. This assumption is the same one that we made in the case of the *NPN* BJT.

The similarities between the waveforms present throughout the *NPN* circuit and the waveforms throughout the *PNP* circuit are considerable. If you understand the operation of these circuits, you will know that these waveforms are correct; and, you should be able to draw the waveforms for either polarity BJT and any signal-waveform input. Note that there is a phase reversal between input-signal voltage and output-signal voltage with either polarity BJT.

A circuit using the universal-bias technique has an emitter resistance. The waveforms associated with the collector circuit are somewhat different from the collector waveforms associated with the grounded emitter case. We will therefore look at the waveforms

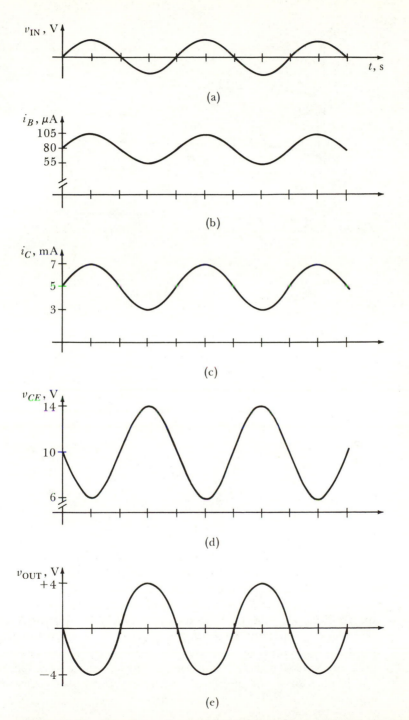

Figure 5.20 Waveforms for an *NPN* BJT amplifier: (a) input voltage; (b) base current; (c) collector current; (d) collector voltage; (e) output voltage.

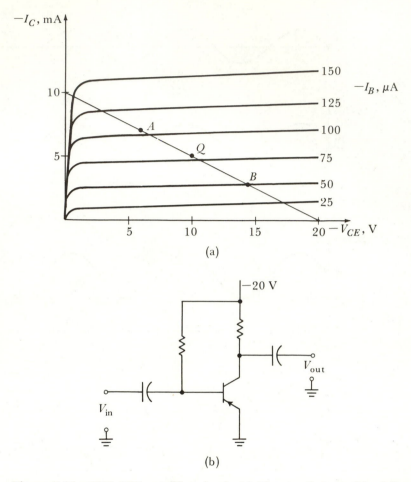

Figure 5.21 *PNP* BJT amplifier circuit. (a) Characteristics and load line. (b) Circuit connection.

involved. As shown in Figure 5.23, we have chosen an input signal that makes use of the same portion of the load line as the preceding examples. We must remember that for the load line shown, R_L plus R_E must be 2 kΩ. The waveforms of interest for this circuit are shown in Figure 5.24. We have shown i_C and i_E as being the same. In actuality they differ by 80 μA at the Q point. The single-subscripted variables v_E and v_C refer to the emitter and collector voltages wrt ground. Note that the emitter voltage v_E is in phase with i_E and i_C. v_{CE} is of course out of phase because of the minus sign in the equation

$$v_{CE} = V_{BB} - i_C(R_L + R_E)$$

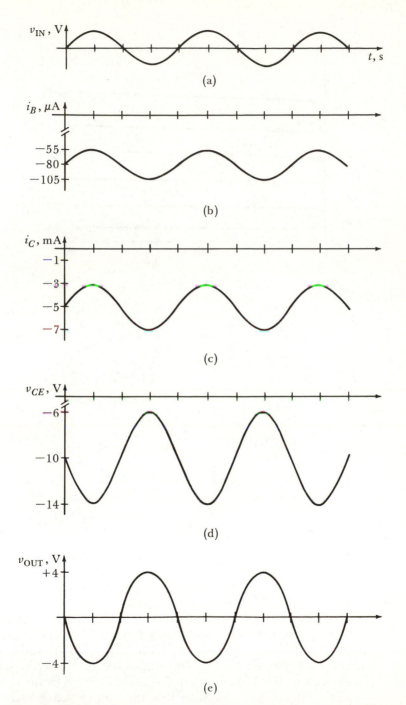

Figure 5.22 Waveforms for a *PNP* BJT amplifier: (a) input voltage; (b) base current; (c) collector current; (d) collector voltage; (e) output voltage.

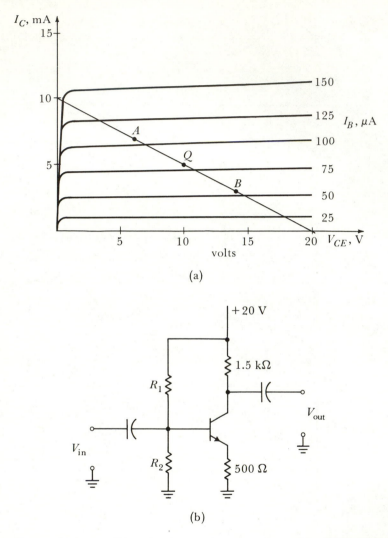

(a)

(b)

Figure 5.23   Amplifier circuit. (a) Characteristics and load line. (b) Circuit connection.

The collector voltage v_C (wrt ground) is the sum of the emitter voltage v_E and the collector-to-emitter voltage v_{CE}. The signals v_E and v_{CE} are out of phase. Since v_{CE} is larger, the signal voltage at the collector is in phase with v_{CE}. The output voltage v_{OUT} is the same as v_C except that the dc component has been removed by the coupling capacitor. It can easily be seen that the output voltage v_{OUT} is of smaller magnitude than v_{CE}.

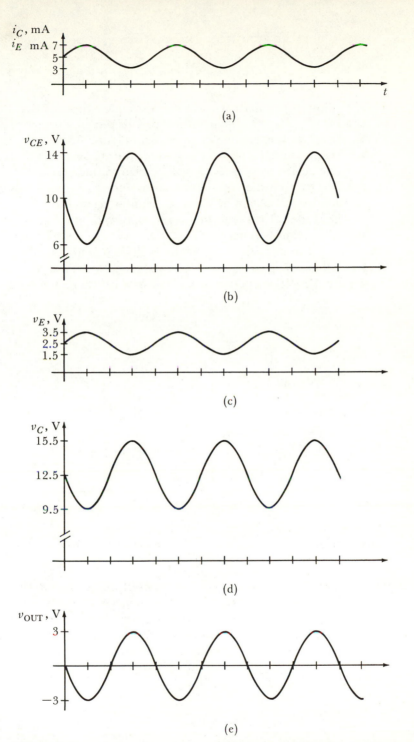

Figure 5.24 Waveforms for *NPN* amplifier circuit with emitter resistor. (a) Collector current or emitter current; (b) collector-to-emitter voltage; (c) emitter-to-ground voltage; (d) collector-to-ground voltage; (e) output voltage.

It should be apparent from our example problems that the output-signal voltage is smaller when part of the total circuit resistance is placed in the emitter circuit as in Figure 5.23(b). In general, adding an emitter resistance will lower the voltage gain of an amplifier circuit. In many cases we will be glad to trade the extra gain we would get from the zero-emitter-resistance case for the extra bias stability we can obtain by using an emitter resistance.

The trade off between dc bias stability and ac voltage gain mentioned previously is not absolute. That is, there is a scheme whereby we can have both desirable features. The scheme involves nothing more than a bypass capacitor across the emitter resistor, as shown in Figure 5.25. The emitter-bypass-capacitance value is chosen so that

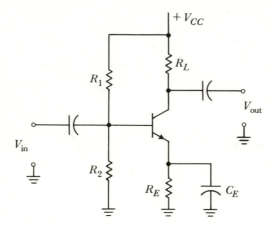

Figure 5.25   Basic-amplifier circuit with emitter-bypass capacitor.

X_c is very low at the signal frequencies of interest. Thus the emitter impedance is near zero at signal frequencies so that voltage gain is large. However, at dc the emitter impedance is R_E, so that we have the good bias stability associated with an emitter resistance. This circuit can be analyzed graphically with an ac load line or analytically by use of ac equivalent circuits. Both methods will be considered later.

5.8   AC LOAD LINES

There are cases where the graphical load-line technique must be applied somewhat differently than we have done so far. Two common cases are when an emitter bypass capacitor is used and when amplifier stages are ac-coupled to a load. In general, if an amplifier stage has an ac load line which differs from the dc load line, both lines must be taken into account for a correct analysis.

Consider the circuit of Figure 5.26(a). It is a simple amplifier circuit with an emitter resistor and an emitter bypass capacitor. The dc load line is drawn in normal fashion. Endpoints are

$$V_{CE} = 20, \quad I_C = 0 \quad \text{and} \quad V_{CE} = 0, \quad I_C = \frac{20 \text{ V}}{4 \text{ k}\Omega} = 5 \text{ mA}$$

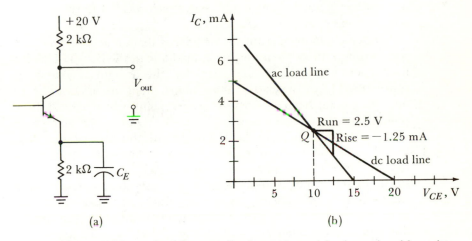

(a)   (b)

Figure 5.26 ac load-line application: (a) partial schematic with emitter bypass capacitor; (b) applicable dc and ac load lines (characteristic curves omitted for clarity).

Since the dc collector-emitter current flows through both collector and emitter resistors, a total of 4 kΩ is used to calculate the current-axis intercept of the dc load line. The dc load line is drawn in Figure 5.26(b). An arbitrarily defined Q point is identified.

Now we want to learn to draw an ac load line. If we make careful note of three facts, we will find the construction of an ac load line to be a relatively easy task.

The first fact we want to make note of is that the slope of a dc load line is the negative reciprocal of the total dc resistance in the collector and emitter circuits. For the example of Figure 5.26, this is expressed as

$$\text{Slope (dc)} = -\frac{1}{R_{\text{collector}} + R_{\text{emitter}}}$$

$$= -\frac{1}{4 \text{ k}\Omega} = -0.25 \text{ mA/V}$$

It is obvious from the construction that the load line has a negative slope. Thus the minus sign. The magnitude of the slope for our example can be calculated from the construction as

$$\text{Slope (dc)} = \frac{\text{rise}}{\text{run}}$$

$$= -\frac{5 \text{ mA}}{20 \text{ V}} = -0.25 \text{ mA/V}$$

This little calculation verifies that the magnitude of the slope of the dc load line is numerically equal to the reciprocal of the sum of the collector and emitter resistances.

The second item we want to consider is that an ac load line must pass through the Q point on the dc load line of the circuit. This fact makes sense if we stop to realize that the ac signals (that is, as represented by the ac load line) are alternations or variations of the operating point from a steady reference point, which we call the Q point.

The third factor to consider has to do with the slope of an ac load line. The ac load line is to represent the response of the circuit to ac signals. Since bypass-capacitor values are chosen to be essentially short circuits at the ac frequencies of interest, all bypassed resistors will be considered as being shorted out for our ac load-line calculations. The slope of an ac load line, then, is the negative reciprocal of the sum of the emitter and collector ac impedances. In the specific example of Figure 5.26 we have,

$$\text{Slope (ac)} = -\frac{1}{Z_{\text{collector}} + Z_{\text{emitter}}}$$

$$= -\frac{1}{2 \text{ k}\Omega + 0} = -0.5 \, (10^{-3})$$

$$= -0.5 \text{ mA/V}$$

The combination of items 2 and 3 gives us enough data so that we can draw an ac load line by way of the point-slope form. Since we have already determined the point and the slope, let us proceed with construction of the ac load line. From the Q point we choose some convenient rise and run values so that the slope is -0.5 mA/V. In this case we arbitrarily choose a run of 2.5 V. Then

$$\text{Slope (ac)} = \frac{\text{rise}}{\text{run}} = -0.5 \text{ mA/V}$$

$$0.5 \, (10^{-3}) = \frac{\Delta I_C}{\Delta V_{CE}} = \frac{\Delta I_C}{2.5}$$

Solving for the rise, we have

Rise $= \Delta I_C = -1.25$ mA

Thus if $\Delta V_{CE} = +2.5$ V, ΔI_C must be -1.25 mA. The construction shown on the load line reflects our chosen rise and run values.

Another common situation where an ac load line is helpful is shown in Figure 5.27. In this case the collector-signal voltage is ac-coupled to a load resistance R_L. This has the effect of lowering the ac impedance in the collector circuit. The load resistance R_L may well represent the equivalent input impedance of a following amplifier stage. (Another method of handling input and output impedances is shown in Section 8.2, and cascading amplifier stages are considered in Section 9.3.)

For our specific example, the ac impedance in the collector circuit is 8 kΩ in parallel with 8 kΩ or an equivalent of 4 kΩ. The 8-kΩ

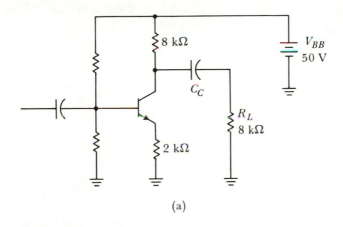

(a)

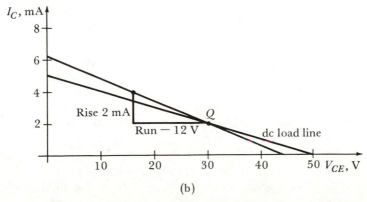

(b)

Figure 5.27   ac-coupled amplifier: (a) circuit; (b) ac and dc load lines.

resistors are in parallel under the specified ac conditions because both the coupling capacitor C_C and the 50-V battery are short circuits at the ac signal frequencies. Thus it is seen that both 8-kΩ resistors are connected from collector to ground. (More will be said about ac equivalent circuits in the next section.)

The sum of the equivalent collector and emitter ac impedances in the circuit of Figure 5.27 is 6 kΩ. Thus we can determine the slope of the ac load line.

$$\text{Slope (ac)} = -\frac{1}{Z_{\text{collector}} + Z_{\text{emitter}}}$$

$$= -\frac{1}{4 \text{ k}\Omega + 2 \text{ k}\Omega} = -\frac{1}{6 \text{ k}\Omega}$$

$$= -0.1667 \text{ mA/V}$$

Now let us determine some convenient rise and run values so that we can plot the ac load line. We arbitrarily choose a rise of $\Delta I_C = 2$ mA. Thus

$$\text{Slope (ac)} = \frac{\text{rise}}{\text{run}} = -0.1667 \text{ mA/V} = \frac{\Delta I_C}{\Delta V_{CE}}$$

$$\Delta I_C = 2 \text{ mA} = (-0.1667)(10^{-3}) \Delta V_{CE}$$

$$\Delta V_{CE} = -\frac{2(10^{-3})}{0.1667(10^{-3})} = -12 \text{ V}$$

The construction shown in Figure 5.27(b) uses our chosen rise and run values to plot the ac load line.

At this point we should consider the usefulness of load lines. They will show us nothing that we cannot find out mathematically. In fact they are no more than a graphical representation of the appropriate mathematics. If load lines are of any value, it is because of one's preference in working with graphical techniques.

Consider the load lines of Figure 5.26(b). At a glance we can see, from the dc load line, that at dc (or low frequencies) the V_{CE} voltage can increase to 20 V before severe distortion (limiting or clipping) occurs. However, it is apparent from the ac load line that at the signal frequencies of interest, V_{CE} can only increase to 15 V before clipping occurs. At signal frequencies the emitter resistance is shorted out by C_E; thus at signal frequencies V_{ce} and V_{out} are equal. We have just determined at a glance from the load line that the maximum signal output of the amplifier is 5 V peak.

5.9 BIPOLAR JUNCTION TRANSISTOR AC EQUIVALENT CIRCUITS

Many times when we are dealing with amplifier circuits we are interested only in ac signals. In many cases the dc levels that are present to bias the BJT are troublesome to us. We usually put coupling capacitors in the circuit to get rid of any dc levels at the input and output terminals. In cases where we are interested only in the ac signals, it seems advantageous, if possible, to draw a simplified circuit. The simplified circuit has meaning *only* when dealing with ac signals.

Let us consider how an amplifier circuit might be simplified when dealing with ac signals exclusively. The coupling capacitors were chosen in value so as to have small X_c at signal frequencies. Therefore, in the ac equivalent circuit, we will replace all capacitors with short circuits. Ideally batteries have no internal resistance or reactance. Batteries are *not* ac voltage sources. We can therefore replace all batteries with short circuits in the ac equivalent circuit. Figure 5.28 shows two forms of the basic amplifier circuit and an ac equivalent circuit of each. Note that every resistance shown in Figure 5.28(b) and (d) has one lead connected to ground (ac ground). Note also that R_1 and R_2 are in parallel and therefore could be shown as one resistor whose resistance is equal to the parallel combination of R_1 and R_2.

The equivalent circuits shown in Figure 5.28(b) and (d) represent what the circuits of Figure 5.28(a) and (c) would look like to you if you were an ac signal. Since the batteries look like short circuits to you (an ac signal), all resistors appear to be connected to ground. You cannot tell whether the BJTs are *NPN* or *PNP*. You cannot tell whether the BJTs are germanium or silicon. If the emitter resistor of Figure 5.28(c) were bypassed with a suitable capacitor, and this is sometimes done, the equivalent circuits of (b) and (d) would look like exactly the same configuration to you. You could tell the difference between these equivalents only if R_1 in parallel with R_2 had a different resistance value than R_B, or if you noticed that $R_L = 2$ kΩ in one case and 1.5 kΩ in the other case.

We should now realize that the equivalent circuits of Figure 5.28 are much simplified, but that we can make one further equivalence. We have so far done nothing with the BJT except to draw the circuit diagram such that the BJT appears to have no bias arrangements. The circuits are not in a suitable form for direct application of loop or nodal analysis.

There are several forms of ac equivalent circuits for BJTs. In the normal lower-frequency ranges, the most widely used equivalents are probably the T equivalent circuit and the parameter form of

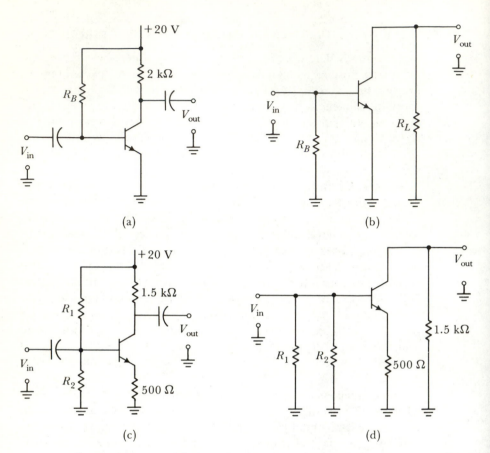

Figure 5.28   ac equivalent circuits of basic amplifiers: (a) *NPN* amplifier circuit with one-battery bias circuit; (b) an ac equivalent circuit of part a; (c) *NPN* amplifier circuit with universal-bias circuit; (d) an ac equivalent circuit of part c.

equivalent circuit. There are BJT ac equivalent circuits for the impedance (Z) parameters, the admittance (Y) parameters, the hybrid (h) parameters and several others. Each equivalent circuit has certain advantages and disadvantages. Most manufacturers' data sheets for transistors will give typical values for the h parameters. The h-parameter equivalent circuit of a transistor is shown in Figure 5.29.

It has been decided that in this book, we will not use any of the standard ac equivalent circuits of transistors that have been listed previously. If your curriculum has a course on BJTs and BJT circuits that follows the course for which this book is used, the standard equivalent circuits will no doubt be covered in detail. If you

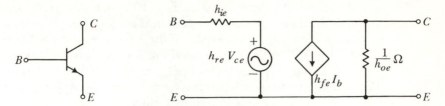

Figure 5.29 h-parameter equivalent circuit of a BJT.

will not be having such a course, it seems advantageous to use a very simple equivalent circuit that will contribute greatly to an intuitive understanding and insight into the operation of transistor amplifier circuits.

The BJT ac equivalent circuit we have chosen to use is shown in Figure 5.30. Note that this equivalent circuit is closely related to the h-parameter equivalent circuit. We want to develop this equivalent

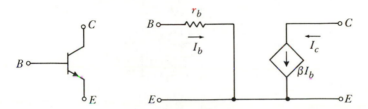

Figure 5.30 BJT and its ac equivalent circuit.

circuit as an outgrowth of an understanding of BJT action. Remember that the emitter-base junction of a BJT is forward-biased. There is therefore a dc turn-on voltage across the junction and some dynamic resistance that describes the ac voltage-current relations. The ac equivalent circuit neglects the dc voltage drop but must, of course, show the ac voltage-current relationship. The emitter-base junction is represented in the ac equivalent circuit as a dynamic resistance r_b where the subscript denotes base. With respect to the collector-base junction, we have BJT action when we have collector current which is a function of base current. In fact, we have previously defined the parameter β such that the ac collector current is $I_c = \beta\, I_b$. Our equivalent circuit shows the collector current to be produced by a dependent current source that produces an ac current magnitude of β times the ac base current. Our chosen simple ac equivalent circuit is thus seen to represent quite closely what we know as BJT action.

It should come as no surprise that the BJT manufacturers' parameter specifications will supply us with numerical values. The r_b and β values will usually be given as part of the h parameters. The dynamic resistance r_b will be given by h_{ie} and the forward-current gain β will be given as h_{fe}. Typical values might be $r_b = 500 \ \Omega$ and $\beta = 50$. These values will vary considerably from one BJT type to another. They will vary from unit to unit in a batch of a given type number whose typical values have been given by the manufacturer.

EXAMPLE 5.1   We will now undertake an example problem to show the usefulness of ac equivalent circuits. All signal quantities are ac. Suppose that we have a circuit as shown in Figure 5.31. The ac

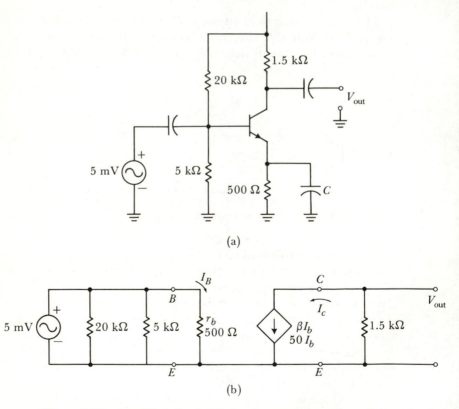

Figure 5.31   Amplifier circuit: (a) circuit connection; (b) equivalent circuit.

equivalent circuit is shown in part b. Note that in the equivalent circuit, the two bias resistors are in parallel with the base resistance. Since we have a voltage-source input, the base current is not affected by the presence of the two-bias resistors. Base current is

$$I_b = \frac{5 \text{ mV}}{500 \text{ }\Omega} = 10 \text{ }\mu\text{A}$$

The output ac signal voltage is

$$V_{\text{out}} = -I_c R_L = -I_c \text{ } 1.5 \text{ k}\Omega$$

$$= -\beta I_b \text{ } 1.5 \text{ k}\Omega$$

$$= -50 \text{ } (10 \text{ }\mu\text{A})(1.5 \text{ k}\Omega) = -0.75 \text{ V}$$

The gain, or amplification, of this example circuit is

$$A_V = \frac{V_{\text{out}}}{V_{\text{in}}} = \frac{-0.75}{5 \text{ mV}} = -150$$

This amplifier has a gain of 150 and a phase shift of 180°. The phase shift is indicated by the minus sign.

EXAMPLE 5.2 Now let us work another example problem. A schematic diagram and an ac equivalent circuit are shown in Figure 5.32. This is a very commonly used circuit, so be sure you understand the problem.

Note that the signal is ac-coupled to a load resistance R_L. This lowers the ac gain somewhat, as we have previously discussed in terms of ac load lines. The output-signal voltage is the voltage across the load resistor R_L and the output-signal current is the ac current through the load resistor.

The first thing that we must calculate is the voltage gain. V_{in} and V_{out} are defined as shown in Figure 5.32.

In order to find V_{out}, we need to know the base current I_b. In Figure 5.32(b) it should be apparent that the value of I_b is independent of the values of R_1 and R_2. We find the value of I_b by writing a loop equation around the loop including V_{in} and r_b. We obtain

$$+V_{\text{in}} - I_b r_b - I_E R_E = 0$$

Fortunately, we can see by inspection that I_e is the sum of I_b and the current from the dependent current source βI_b. Thus,

$$V_{\text{in}} = I_b r_b + (\beta I_b + I_b) R_E$$

$$= I_b [r_b + (\beta + 1) R_E]$$

or

$$I_b = \frac{V_{\text{in}}}{r_b + (\beta + 1) R_E} = \frac{V_{\text{in}}}{500 + (51)(510)}$$

$$= \frac{V_{\text{in}}}{26.510 \text{ k}\Omega}$$

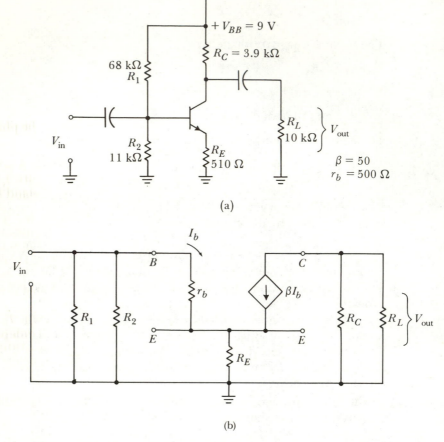

(a)

(b)

Figure 5.32 Circuit for Example 5.2. (a) Schematic diagram; (b) ac equivalent circuit.

Note that the current through the dependent current-source divides between R_L and R_C. Using the current division rule, we obtain

$$I_{rl} = \beta I_b \frac{R_C}{R_L + R_C} = 50 \frac{V_{in}}{26,510} \frac{3.9 \text{ k}\Omega}{(3.9 + 10) \text{ k}\Omega}$$

$$= 0.529 V_{in} \text{ mA}$$

The signal voltage across R_L is

$$V_{out} = I_{rl}R_L$$

$$= (0.529)(10^{-3})(V_{in})(10)(10^3)$$

$$= 5.29 V_{in}$$

Rearranging the last equation, we find that the voltage gain is

$$A_V = \frac{V_{out}}{V_{in}} = 5.29$$

An alternate way of finding the output-signal voltage would have been to note that the output voltage is taken across the parallel combination of R_L and R_C. Since we know that the current βI_b is through the parallel combination, it is a simple multiplication to find the voltage across the parallel combination.

As a second part of this example, let us find the current gain of the amplifier. Input current is the signal current supplied by the signal source. Output current is the signal current in the load resistor R_L. We found the current in the load resistor R_L in the first part of this example, so our main task here is to find the signal current supplied by the signal source.

Referring to Figure 5.32(b), we see that the signal source supplies current to R_1, R_2, and the base of the transistor. We have previously found I_b, so let us proceed to find I_{r1} and I_{r2}. By Ohm's law,

$$I_{r1} = \frac{V_{in}}{R_1} = \frac{V_{in}}{68 \text{ k}\Omega} \quad \text{and} \quad I_{r2} = \frac{V_{in}}{R_2} = \frac{V_{in}}{11 \text{ k}\Omega}$$

Returning to the first part of this example to get the value of I_b, we can write the equation for signal-input current I_{in} as,

$$I_{in} = I_{r1} + I_{r2} + I_b$$

$$= \frac{V_{in}}{68 \text{ k}\Omega} + \frac{V_{in}}{11 \text{ k}\Omega} + \frac{V_{in}}{26.51 \text{ k}\Omega}$$

$$= V_{in} \left(\frac{1}{68 \text{ k}\Omega} + \frac{1}{11 \text{ k}\Omega} + \frac{1}{26.51 \text{ k}\Omega} \right)$$

$$= V_{in} (0.1433)(10^{-3})$$

Now that we know both I_{in} and I_{r1}, we can easily find the current gain as

$$A_I = \frac{I_{out}}{I_{in}} = \frac{I_{rl}}{I_{in}} = \frac{0.529 \ V_{in} \text{ mA}}{0.1433 \ V_{in} \text{ mA}}$$

$$= 3.69$$

The current gain just calculated seems low in view of the fact that the transistor has a current gain of 50. However, only about 25% of the input current is transistor base current. Also note that owing to current-divider action, less than one-third of the transistor collector current is delivered to the load resistor.

EXAMPLE 5.3   This example also refers to Figure 5.32. In this case, we want to calculate the input impedance and the output impedance of the circuit. The input impedance is the load seen by the signal-input voltage-source V_{in}. The output impedance of the amplifier circuit is the Thévenin equivalent resistance of the circuit driving the load resistance R_L.

The input impedance of the circuit can be calculated in a straight-forward manner. First we determine the amount of current supplied by the signal-voltage source. Then we determine the input imped-ance, that is, the equivalent impedance across which the signal source is connected, as the ratio of signal-input voltage to signal-input current.

In Figure 5.32(b) we see that the signal voltage-source supplies three separate currents.

$$I_{in} = I_{r1} + I_{r2} + I_b$$

In Example 5.2 we found this value to be

$$I_{in} = \frac{V_{in}}{26.51 \text{ k}\Omega}$$

Rearranging the equation, we obtain the input impedance as

$$Z_{in} = \frac{V_{in}}{I_{in}} = 26.51 \text{ k}\Omega$$

To calculate the output impedance, we depend on the superposi-tion idea that the total response of a system is the sum of the responses of the system to each source acting separately. In this case, we are interested in the response at the output terminals, so we disable all existing independent signal sources and connect a source across the output terminals. In Figure 5.33(a) we see the circuit after V_{in} was disabled and V_1 was installed across the (former) output terminals. We note that by disabling the signal-input voltage-source, we have placed a short across R_1 and R_2. In Figure 5.33(b), we have redrawn the circuit in a simpler form and defined some current loops. The net effect of what we have done is to consider the (former) output terminals as input terminals for which we plan to find the input (formerly output) impedance.

We can get all the information we need about the circuit of Figure 5.33(b) by writing two independent loop equations. One loop includes the V_1 source and R_C. The other loop includes r_b and R_E. The resulting equations are

$$V_1 - I_1 R_C + \beta I_b R_C = 0 \text{ or } V_1 = R_C(I_1 - \beta I_b)$$

and $r_b I_b + R_E(\beta + 1)I_b = 0$

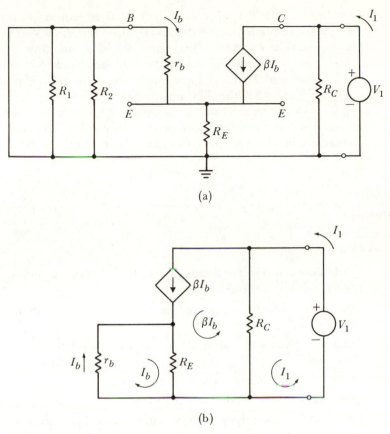

(a)

(b)

Figure 5.33   ac equivalent circuits for Example 5.3: (a) original configuration; (b) redrawn configuration defining currents.

Solving the base loop equation we obtain,

$$I_b [r_b + (\beta + 1)R_E] = 0 \quad \text{or} \quad I_b = 0$$

Substituting for I_b in the first equation we obtain

$$V_1 = R_C I_1 \quad \text{or} \quad R_C = \frac{V_1}{I_1} = 3.900 \text{ k}\Omega$$

Thus we have determined that the input impedance of the (former) output terminals is R_C. Stated in a manner consistent with our original goals, we have

$$Z_{\text{out}} = \frac{V_1}{I_1} = R_C = 3.900 \text{ k}\Omega$$

A slightly different approach than we used would be to place a current source I_1 across the (former) output terminals and solve for the appropriate V_1 value. The results would be the same.

If you were somewhat surprised when the calculated value of I_b for the circuit of Figure 5.33(b) turned out to be zero, consider the circuit of Figure 5.32(b). There we see that I_b is one of the currents supplied by the signal source V_{in}. As V_{in} goes to zero, I_b will go to zero and so will the dependent current source βI_b. Thus in addition to mathematical correctness, the result is intuitively correct.

EXERCISES

QUESTIONS

Q5.1 How is it possible to get current flow through a reverse-biased collector-base junction?

Q5.2 What differences should be noted between germanium and silicon BJTs?

Q5.3 In what ways are *NPN* transistors and *PNP* transistors different?

Q5.4 What is meant by the Q point?

Q5.5 Why is bias needed in transistors?

Q5.6 What causes the minus sign in the load-line equation?

Q5.7 What is a load line?

Q5.8 Why do some circuits have an ac load line different from the dc load line?

Q5.9 Why must the Q point of an amplifier always be located on the dc load line?

Q5.10 What are some factors influencing the choice of an optimum Q point?

Q5.11 How does the universal-bias circuit achieve a self-adjusting feature?

Q5.12 What are the advantages and limitations when working with ac equivalent circuits?

PROBLEMS

P5.1 Draw the schematic symbol for an *NPN* transistor. Show current directions and battery-voltage polarities needed.

P5.2 Repeat Problem 5.1 for a *PNP* transistor.

P5.3 Graphically determine the value of β for a BJT near the point where

$$V_{CE} = 15 \text{ V}, \qquad I_C = 5 \text{ mA}$$

The BJT has characteristic curve *A* shown in Figure 5.34.

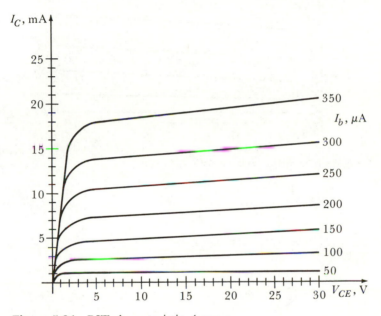

Figure 5.34 BJT characteristic *A* curve.

P5.4 Repeat Problem 5.3 where

$$V_{CE} = 15 \text{ V}, \qquad I_C = 12 \text{ mA}$$

P5.5 Calculate the value of α for the BJT of Problem 5.3.

P5.6 Calculate the value of α for the BJT and operating point specified in Problem 5.4.

P5.7 Find the value of β for a BJT when

$$V_{CE} = 10 \text{ V}, \qquad I_C = 0.5 \text{ mA}$$

The BJT has characteristic *B* shown in Figure 5.35.

P5.8 Repeat Problem 5.7 near the point where

$$I_B = 300 \text{ } \mu\text{A}, \qquad I_C = 9 \text{ mA}$$

P5.9 Draw the load line for the *NPN*-transistor circuit shown in Figure 5.36. The BJT has characteristic *A* shown in Figure 5.34.

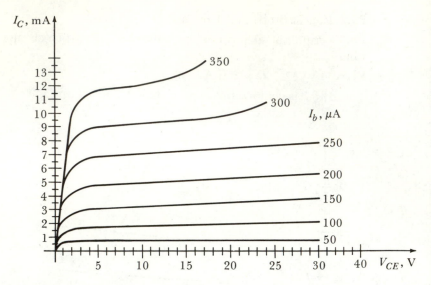

Figure 5.35   BJT characteristic *B* curve.

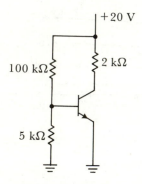

Figure 5.36   See Problem 5.9.

P5.10 Repeat Problem 5.9 where the transistor has characteristic curve *B* of Figure 5.35.

P5.11 Construct a load line for the *PNP*-transistor circuit shown in Figure 5.37. The transistor has characteristic *A* shown in Figure 5.34.

P5.12 Repeat Problem 5.11 when the transistor characteristic curves are as shown in Figure 5.35.

P5.13 Draw dc and ac load lines for the circuit shown in Figure 5.38. The transistor has characteristic *B* of Figure 5.35.

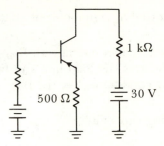

Figure 5.37 See Problem 5.11.

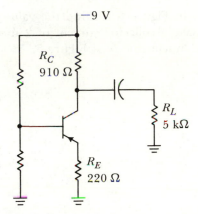

Figure 5.38 See Problem 5.13.

P5.14 Draw the dc and ac load lines for the circuit of Figure 5.39 when the transistor has characteristic *B* of Figure 5.35.

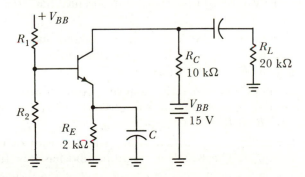

Figure 5.39 See Problem 5.14.

P5.15 Find the maximum undistorted ac signal-output voltage for the circuit of Problem 5.13.

P5.16 Assume that we have a BJT amplifier circuit with $V_{BB} = 25$ V and a collector resistor of 2.5 kΩ. Draw the load line and find the optimum Q point for the following conditions. Use the BJT characteristic A of Figure 5.34.

    (a) Maximum signal-handling capability.

    (b) Best linearity (fidelity).

    (c) Minimum dc-power dissipation.

P5.17 Repeat Problem 5.16 for BJT characteristic B.

P5.18 For the circuit shown in Figure 5.40, find the value of R_B needed to bias the BJT to have a collector current of 2 mA. The transistor is silicon and has characteristic B of Figure 5.35.

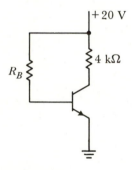

Figure 5.40   See Problem 5.18.

P5.19 Find the value of R_B needed to bias the BJT circuit of Figure 5.41 at $V_{CE} = -15$ V. The BJT is germanium and has characteristic A of Figure 5.34.

P5.20 The circuit of Figure 5.42 is to be biased such that the collector current is 10 mA. The transistor is silicon and has $\beta = 50$. Let $I_{R2} = 5I_B$. Find R_1 and R_2 values.

P5.21 The circuit of Figure 5.42 has a germanium transistor that is to be biased at $V_{CE} = 9$ V (not V_C wrt ground). Find the needed values of R_1 and R_2 if $\beta = 100$ and when $I_{R2} = 2I_B$.

P5.22 Find the value of R_B needed to bias the transistor of Figure 5.43 at $V_{CE} = -20$ V. The BJT is germanium and has $\beta = 100$.

P5.23 Repeat Problem 5.22 if the BJT is silicon and is to be biased at $I_C = 15$ mA. The value of β is 150.

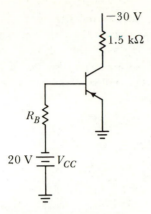

Figure 5.41   See Problem 5.19.

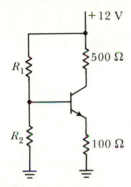

Figure 5.42   See Problems 5.20 and 5.21.

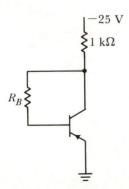

Figure 5.43   See Problem 5.22.

P5.24 Use graphical techniques to determine the current gain of the circuit shown in Figure 5.44. The ac current-source provides the input signal current. The ac current through the collector resistance is considered the output current. Assume that the BJT is biased at I_B = 150 μA and that the input-signal current is 50 μA peak. The BJT has characteristic A of Figure 5.34.

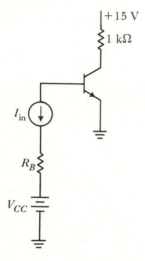

Figure 5.44   See Problem 5.24.

P5.25 Use an ac equivalent circuit to find the voltage gain of the circuit shown in Figure 5.45. The BJT parameters are β = 80 and r_b = 1000 Ω.

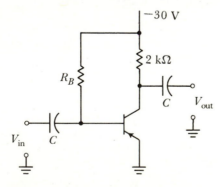

Figure 5.45   See Problem 5.25.

P5.26 A frequently used amplifier circuit is shown in Figure 5.46. The BJT is characterized by $\beta = 150$ and $r_b = 1500\ \Omega$. Use an ac equivalent circuit to find the voltage gain of the circuit. Note that R_1 and R_2 values are not given or needed.

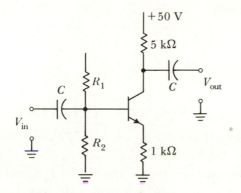

Figure 5.46   See Problem 5.26.

P5.27 Find the ac current gain for the circuit shown in Figure 5.47. Input current is the current delivered to the V_{in} terminals. Output-signal current is the signal current through the load resistor R_L. The BJT is specified as $\beta = 20$ and $r_b = 500\ \Omega$.

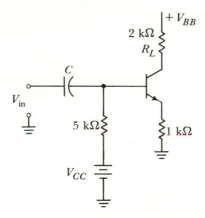

Figure 5.47   See Problem 5.27.

P5.28 Find the ac voltage gain for the circuit shown in Figure 5.48. The BJT has specifications of $\beta = 50$ and $r_b = 800\ \Omega$.

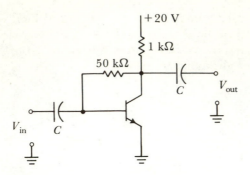

Figure 5.48   See Problem 5.28.

P5.29 Find the input impedance for the circuit of Figure 5.45 when $\beta = 100$, $r_b = 1000 \ \Omega$, and $R_B = 270 \ \text{k}\Omega$.

P5.30 Find the output impedance for the situation of Problem 5.29.

P5.31 For the circuit of Figure 5.46, $\beta = 200$, $r_b = 1000 \ \Omega$, $R_1 = 560 \ \text{k}\Omega$, and $R_2 = 110 \ \text{k}\Omega$. Find the input impedance of the circuit.

P5.32 Calculate the output impedance for the circuit of Problem 5.31.

CHAPTER 6
FIELD-EFFECT TRANSISTORS

Field-effect transistors (FETs) are relative newcomers on the electronics scene. At least they are new in terms of wide availability and usage as compared with the vacuum tube and the BJT.

Great strides have been made in recent years in the development of FETs and further research and development is expected. Much of the present effort has to do with the utilization of FETs in integrated circuits (ICs).

It is the purpose of this chapter to explain the operation of FET devices, so that the reader can analyze or design circuits containing FETs as well as understand the internal components and operation of integrated circuits.

6.1 CONSTRUCTION OF FIELD-EFFECT TRANSISTORS

There are two basic types of field-effect transistor: the junction field-effect transistor and the insulated-gate field-effect transistor. The junction-gate version is known as the JFET or simply FET. The insulated-gate versions are called IGFETs or MOSFETs. The MOS stands for metal oxide semi-conductor.

The fundamental operating principle of field-effect transistors, in either the junction-gate or insulated-gate forms, is quite easy to explain. Conducting leads are connected to a so-called channel of a semiconducting material. The semiconductor is usually silicon. The

207

resistance (or conductance) of the channel is made variable by the proper application of a control voltage. The electric field set up by the control voltage is responsible for varying the resistance of the channel. The exact mechanism is somewhat different for the IGFET and JFET cases. These differences will be apparent as we discuss each type separately.

A field-effect transistor is basically a three-terminal device. Each terminal has a distinct name. The terminals are named drain, source, and gate. The endpoints of the semiconductor channel are called the *drain* and the *source*. The control element is called the *gate*. Control voltages are applied between the gate and the source.

A junction FET may be constructed something like that shown in Figure 6.1. The physical shape shown is not unique. For example, the channel could be cylindrical in shape with the gate being an annular ring of opposite polarity material. Many other shapes are possible. Construction is not necessarily symmetrical as shown.

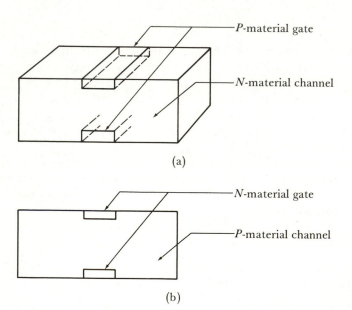

(a)

(b)

Figure 6.1 Junction FET construction: (a) oblique view of *N*-channel JFET; (b) side view of *P*-channel JFET.

In Figure 6.2 we have pictorials and associated schematic symbols for JFETs showing the name of each element. Batteries are shown in Figure 6.2 to indicate the necessary voltage polarities for proper operation.

We sometimes say that the channel resistance is a function of gate

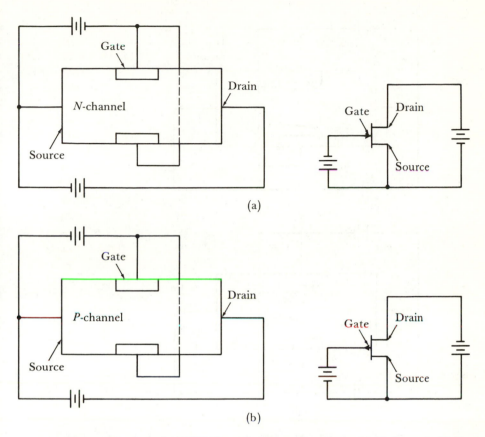

Figure 6.2 Junction FET bias polarities and symbols: (a) N channel; (b) P channel.

voltage. In reality it is the voltage between gate and source that controls the channel resistance. In order to keep the gate-to-channel PN junction reverse-biased, the gate wrt source voltage is always of opposite polarity relative to the drain wrt source voltage. (See Figure 6.2.) Figure 6.3 is an attempt to show how the effective channel width, and thus the channel resistance, varies with the magnitude of the gate wrt source voltage.

In Chapter 2 we saw that for a nonconducting PN junction, there is a region near the junction that is devoid of unbound charges (electrons or holes). This region is known as the depletion region. The width of the depletion region is dependent on the magnitude of the reverse bias across the junction. The depletion region shown for the field-effect transistor in Figure 6.3 is exactly the same as that discussed in Chapter 2. It is produced by a reverse-biased PN junction. Since the depletion region is devoid of unbound charge

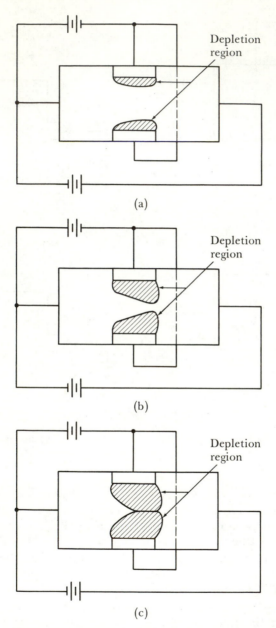

Figure 6.3 JFET channel conditions as gate wrt source voltage is varied. The gate wrt source voltage is increasing from part a through part c.

carriers, we can see in Figure 6.3 that the effective (that is, conductive) width of the channel decreases as the gate wrt source voltage increases the reverse bias on the *PN* junction. Thus the gate wrt source voltage does in fact control the conductivity of the channel. Consider the case shown in Figure 6.3(c). The gate wrt source voltage is of such magnitude as to cause the depletion regions to meet. The effective channel width is zero. There is no conductive channel. Under these circumstances, the FET essentially has infinite drain-to-source resistance. Since it is nonconductive, the FET is said to be in the *cutoff*, or *pinch-off, state.*

There is a problem with the pinch-off terminology. The word pinch off has another separate and distinct meaning. When referring to the cutoff-type of pinch-off voltage, we will use the symbol $V_{P(GS)}$. This is to be interpreted as the voltage between gate and source which causes channel current to go to zero at a specified drain wrt source voltage.

In Figure 6.3, the depletion region is not symmetrical in shape around the gate region. Consider the FET shown in Figure 6.4. It has only one gate region for simplicity. Assume 10 V is connected from drain to source as shown. If the channel is of uniform dimensions and its effective dimensions are changed a negligible amount by the depletion region around the gate, then the voltage distribution at various points along the channel length, with respect to the source, will be as shown. The gate wrt channel voltage is shown as 8 V near one end of the gate and as 4 V at the other end of the gate. This is a reverse-bias voltage. The depletion region is, as expected,

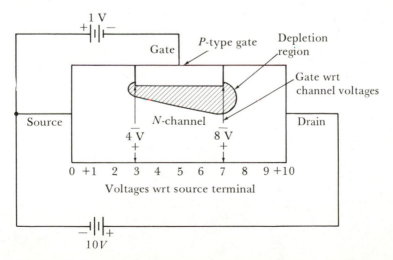

Figure 6.4 Various potentials throughout the structure of a simple JFET.

widest where the reverse bias is highest. Some of the assumptions we
have made in this discussion are not exactly true. Hopefully, however, the shape we have shown for the depletion region of a JFET is
believable.

Consider a field-effect transistor connected as in Figure 6.5. Note
that the gate lead is shorted to the source lead, that is, $V_{GS} = 0$. As
the V_{DS} voltage is increased from zero, we would expect some
channel (drain) current. This is true, as indicated in Figure 6.5(b). As
we continue to increase V_{DS} we will get increased reverse bias on the
gate-to-channel junction, increased depletion areas, and a decrease
in effective (that is, conductive) channel dimensions. (Refer to Figure 6.4 if you have trouble visualizing how the gate-channel junction
can be reverse-biased even when the gate wrt source voltage is zero.)
These changes will cause the channel (drain) current to increase a
lesser amount than might be expected. In fact, a point is soon
reached where the channel (drain) current will (essentially) increase

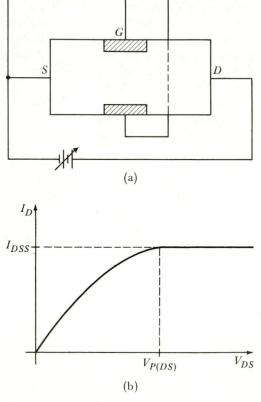

Figure 6.5 (a) Circuit connection and (b) *I-V* characteristic that define the
quantities $V_{P(DS)}$ and I_{DSS} for an FET.

no further. The limiting current is designated as I_{DSS}. This three-subscript terminology means the limiting *d*rain-to-*s*ource current with the other element (in this case, the gate) *s*horted. The drain wrt source voltage that must be applied before I_{DSS} occurs is called the pinch-off voltage. We will call this pinch-off voltage the drain wrt source pinch-off voltage and will use the symbol $V_{P(DS)}$ to minimize any confusion with the previously defined pinch-off quantity.

A method of construction somewhat different from that for a JFET results in a field-effect transistor known as an insulated-gate type (IGFET) or as a metal-oxide-semiconductor type (MOSFET). In Figure 6.6 we have a somewhat simplified version of the constructional details. The gate is insulated from the other elements by a near-perfect insulating layer of silicon dioxide. The metal film of the gate electrode forms one plate of a capacitor. The other plate is the substrate material. As a gate voltage is applied, (that is, a voltage impressed on the gate-to-substrate capacitance), induced charges appear in the channel and substrate regions. Depending on the polarity of the gate voltage, the induced charges may be positive or negative. The induced charges will change the conductivity, or resistivity, of the channel. Whether the conductivity increases or decreases depends on the polarity of the applied gate voltage and whether the device is a *P*-channel or *N*-channel type. A positive gate voltage will increase the conductivity (decrease the resistivity) of an *N*-channel device and decrease the conductivity of a *P*-channel device.

In addition to the *N*-channel and *P*-channel variations of the MOSFET, the MOSFET can be made to operate in either a depletion mode or an enhancement mode. Whether an FET device is considered a depletion device or an enhancement device is primarily determined by the depth of the channel, as shown in Figure 6.6.

In a depletion-type IGFET, the channel dimensions are such that

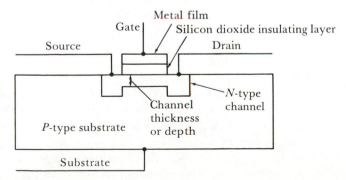

Figure 6.6 Construction of insulated-gate FET (IGFET or MOSFET).

with no bias (that is, no gate wrt source dc voltage), the conductivity of the channel is rather high. The depletion-type device, therefore, may be referred to as a normally-on device. The gate wrt source voltage has control over channel current in this type of device by decreasing the conductivity of the channel.

An IGFET device can be made that operates in an enhancement mode. In this case the channel depth is very small, or nonexistent. Under these conditions we find that the drain current is very small when no gate wrt source voltage is present. For this reason we may refer to enhancement-mode devices as being of the normally-off type. The gate wrt source voltage has control over the drain current in this type of device by increasing the conductivity (that is, enhancing conductivity) of the channel.

In order to understand how the gate voltage is able to control channel conductivity in IGFETs, let us refresh our memory about some properties of capacitors. Figure 6.7 shows schematic symbols of two capacitors. One capacitor is uncharged and the other is charged. The polarity signs on the plates of the charged capacitor

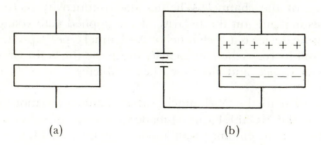

(a)   (b)

Figure 6.7 Parallel-plate capacitors: (a) uncharged; (b) charged.

represent the electric charges that have moved to or from the plates during the charging process. In order to relate this phenomenon to semiconductor terminology, let us consider that the battery voltage applied to the capacitor of Figure 6.7(b) has caused the lower plate to have an excess number of electrons and the upper plate to have an excess number of holes. These charges congregate near the junction with the insulating material.

Now let us return to our discussion of IGFETs. In particular we want to consider the channel-current control mechanism of the gate. In Figure 6.8 we have a *P*-channel enhancement-mode IGFET. Note that the channel is nonexistent between the drain and source electrodes under the condition shown where the voltage between the gate and substrate is zero. This ensures that the device must be of

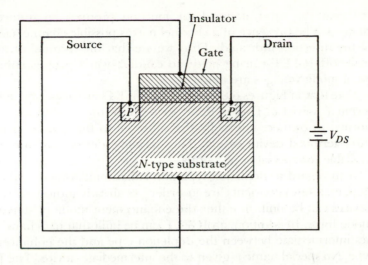

Figure 6.8 *P*-channel enhancement-mode IGFET.

the normally-off type. If this device is to be a useful FET, it must necessarily be an enhancement-mode FET.

When voltages of the correct polarity are applied to the IGFET of Figure 6.8, the situation shown in Figure 6.9 will exist. The gate and the substrate material are, in effect, the plates of a capacitor. The charge indicated on the plates of the capacitor is the result of the gate-substrate voltage $V_{G(sub)}$. The excess holes in the substrate mate-

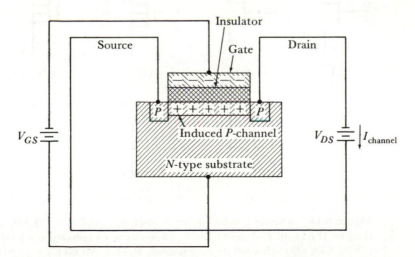

Figure 6.9 *P*-channel enhancement-mode IGFET emphasizing the induced channel.

rial near the substrate-insulator junction produce an induced P channel. The presence of a channel makes possible channel current of the direction indicated. Thus we see that the channel of a normally-off IGFET can be made to conduct when proper enhancement-mode voltages are applied.

Note that in Figures 6.8 and 6.9 the IGFETs are displayed as four-terminal devices. This is done to simplify the explanation. It is normal to connect internally the substrate to the source so that a three-terminal device results. Four-terminal devices, as shown, are available for special applications.

With regard to depletion-type and enhancement-type field-effect devices, a few comments are in order. As already indicated, IGFET devices can be built in either the enhancement-mode or depletion-mode form. In addition, an IGFET can be built that has characteristics intermediate between the depletion type and the enhancement type. No special name is given to the intermediate device. The JFET is exclusively a depletion-mode device. Any attempt to operate a JFET in an enhancement mode to any significant extent will result in forward conduction of the gate-to-channel junction. Such a situation is not normally useful.

The schematic symbols we will be using for the various types of FETs are shown in Figure 6.10. Note that the insulated-gate versions

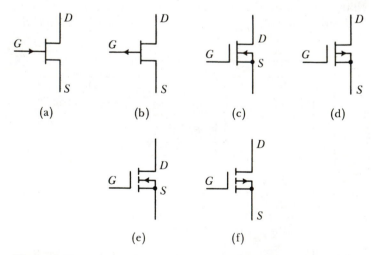

Figure 6.10 Schematic symbols for field-effect transistors (FETs): (a) N-channel JFET; (b) P-channel JFET; (c) N-channel depletion-mode IGFET (MOSFET); (d) P-channel depletion-mode IGFET (MOSFET); (e) N-channel enhancement-mode IGFET (MOSFET); (f) P-channel enhancement-mode IGFET (MOSFET).

are represented by symbols that depict the insulating layer between gate and channel. Note also that the enhancement-mode-device symbols have a broken-channel representation. This is in close agreement with our earlier explanation that there is no complete channel until or unless enhanced by the gate wrt source voltage.

The supply voltages required by an FET differ according to the type of FET used. In Figure 6.11 we have a diagram for each type of

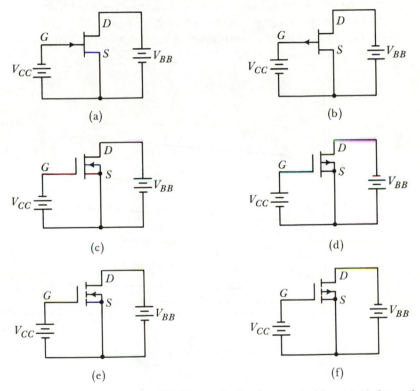

Figure 6.11 Illustration of FET required voltage polarities: (a) *N*-channel JFET; (b) *P*-channel JFET; (c) *N*-channel depletion-mode IGFET (MOSFET); (d) *P*-channel depletion-mode IGFET (MOSFET); (e) *N*-channel enhancement-mode IGFET (MOSFET); (f) *P*-channel enhancement-mode IGFET (MOSFET).

FET showing the correct supply voltages. Enhancement-mode devices have the distinction, and advantage, of requiring the same polarity for drain and gate supply voltages.

6.2 *I-V* CHARACTERISTICS OF FETS

The *I-V* terminal characteristics of a three-terminal device may be displayed on a two-dimensional graph if we let one variable be what is known as a *running variable*. The end result of this technique is the generation of a family of curves. The commonly accepted choices of variables for FETs will display drain current versus drain wrt source voltage. The family of curves comes about by letting the gate wrt source voltage take on several different constant values. The gate wrt source voltage is the running variable. The graph shown in Figure 6.12 is the commonly accepted way of displaying the *I-V* terminal characteristics of FETs.

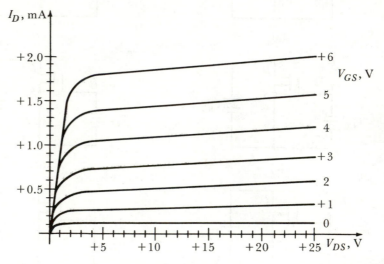

Figure 6.12 *I-V* characteristics of an *N*-channel enhancement-mode MOSFET.

Sometimes a third subscript is used when speaking of three-terminal devices. The third subscript symbol will be either O for open circuit or S for short circuit. The open or short referred to by the third subscript has to do with the element not referred to by the first two subscripts. As examples we say that I_{DSS} means the drain-to-source current when the gate is shorted to the source and that I_{DSO} means the drain-to-source current when the gate is open-circuited.

Now that we understand the usage of subscripted variables, let us return to the *I-V* characteristics of FETs. In particular we want to see how the characteristics differ for depletion-mode operation as compared with enhancement-mode operation. In Figure 6.13 are shown the various characteristics.

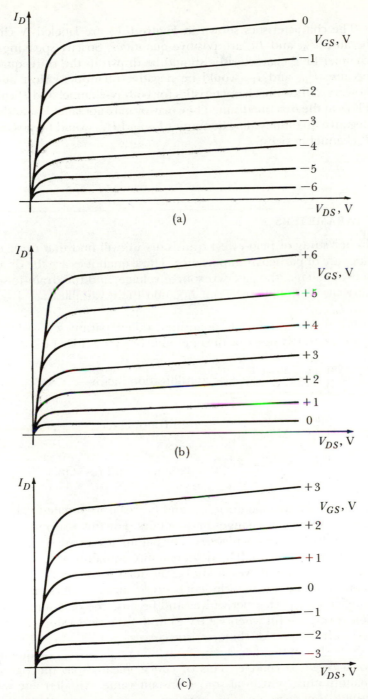

Figure 6.13 *N*-channel FET characteristics: (a) depletion mode (JFET or MOSFET); (b) enhancement mode (MOSFET); (c) Depletion-enhancement mode (MOSFET).

The characteristics shown in Figure 6.13 are labeled N channel because V_{DS} and I_{DS} are positive quantities. Strictly speaking, a P-channel FET characteristic should be drawn in the third quadrant because V_{DS} and I_{DS} would be negative quantities. Most authors, however, draw the characteristics for both N-channel and P-channel FETs in the first quadrant. They may or may not label V_{DS} and I_{DS} as negative quantities. The quantities I_{SD} and V_{SD} would be positive for P-channel devices.

6.3   FET PARAMETERS

In our study of field-effect transistors we will find that three quantities are of particular importance. These quantities are the drain wrt source voltage, the gate wrt source voltage, and the drain-to-source current. In double-subscript notation these variables are V_{DS}, V_{GS}, and $I_{DS} = I_D$.

We define three new quantities, called parameters, which will quite accurately describe the operation of a particular FET device.

$$\mu = \frac{\partial V_{DS}}{\partial V_{GS}} \approx \frac{\Delta V_{DS}}{\Delta V_{GS}}\bigg|_{\Delta I_D = 0} = \text{amplification factor}$$

$$g_m = \frac{\partial I_D}{\partial V_{GS}} \approx \frac{\Delta I_D}{\Delta V_{GS}}\bigg|_{\Delta V_{DS} = 0} = \text{transconductance}$$

$$r_d = \frac{\partial V_{DS}}{\partial I_D} \approx \frac{\Delta V_{DS}}{\Delta I_D}\bigg|_{\Delta V_{GS} = 0} = \text{dynamic drain resistance}$$

Note that the parameters μ, g_m, and r_d are approximated in terms of delta quantities or changes in quantities. For this reason μ, g_m, and r_d are called *dynamic parameters*. The parameters are defined in terms of partial derivatives. The delta-quantity equations will yield nearly the same results if we choose small delta increments. The delta-quantity definition is included because some readers may not be familiar with partial derivatives and because the delta-quantity version is very useful when we try to find numerical values for μ, g_m, and r_d from a set of characteristic curves.

Note that in the delta-quantity defining equations, one quantity is not allowed to change. This of course means that the particular quantity must remain at some constant value. An alternate way of writing the delta-quantity equation is as shown in the following equation for μ.

$$\mu = \frac{\Delta V_{DS}}{\Delta V_{GS}}\bigg|_{I_D=K}$$

The symbol K of course stands for constant value. The other equations may be written in a similar manner.

We will now look at the meaning of each of the defined parameters in an attempt to understand the reason for defining the parameters as they are and to see whether or not the name for each parameter is appropriate. We will also learn to find the numerical value of each parameter from a set of characteristic curves.

The dynamic parameter we have defined with the symbol of the Greek letter μ is called the *amplification factor*. The most common connection for an FET as an amplifier is to apply the input-signal voltage between gate and source terminals and to take the output voltage between the drain and source terminals. Since the parameter μ is defined in terms of the voltage at the same terminal pairs as used in an amplifier circuit, the name of amplification factor seems appropriate. It must be emphasized that amplification factor refers to the operating characteristics of the FET alone and *not* to the amplification of a complete circuit that contains an FET. For example, an FET with an amplification factor of 20 might be used in a circuit whose amplification is 5 or 2 or even ½. The amplification of a simple single FET amplifier circuit can never exceed the amplification factor of the FET used in the circuit.

In order to determine the numerical value of μ for a particular device, we need some characteristic curves for the device. In Figure 6.14 we have a set of characteristic curves with the proper construction for determining the value of μ. The definition of μ is

$$\mu = \frac{\Delta V_{DS}}{\Delta V_{GS}}\bigg|_{\Delta I_D=0 \text{ or } I_D=K}$$

The construction shown does in fact keep drain current constant at 1.5 mA. The values for ΔV_{DS} and ΔV_{GS} must be read from the curves. For this particular construction, $\Delta V_{DS} = 10$ V. The value of ΔV_{GS} is quite difficult to read accurately in this case. By interpolation and a little guesswork we find ΔV_{GS} to be about 50 mV.

$$\mu = \frac{\Delta V_{DS}}{\Delta V_{GS}} = \frac{15 - 5}{0.2 - 0.15}\bigg|_{I_D=1.5 \text{ mA}}$$

$$= \frac{10 \text{ V}}{50 \text{ mV}} = 200$$

Since μ is a ratio of voltages, we see that the amplification factor is

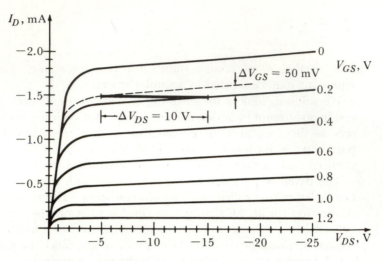

Figure 6.14   Construction for determination of μ value.

dimensionless. The amplification factor of a device in reality tells us how much more effect the gate voltage has, as compared with the source voltage, in controlling the drain current. The drain current must be held constant during this calculation, but if we move from the left end to the right end of the construction line in Figure 6.14, we see that V_{GS} changes in such a direction as to tend to decrease drain current while V_{DS} changes in such a direction as to tend to increase drain current. The fact that the drain current remained constant while V_{DS} changed by 10 V and V_{GS} changed by 0.05 V indicates that the gate voltage has 200 times more control over drain current than does the drain voltage.

The dynamic parameter g_m is called the *dynamic transconductance.* The conductance part of the name refers to the fact that g_m is a ratio of current to voltage. This is the standard definition of conductance. The prefix *trans-* means across the device from input to output. The quantity I_D is considered an output quantity and V_{GS} is considered an input quantity. The name transconductance is seen to be appropriate for the parameter whose symbol is g_m. The quantity g_m is thus an across-the-device, or transfer, conductance.

In order to find a numerical value for g_m, let us look once again at the mathematical definition.

$$g_m = \frac{\Delta I_D}{\Delta V_{GS}}\bigg|_{\Delta V_{DS}=0 \ \text{or} \ V_{DS}=K}$$

Note that V_{DS} must be held constant in order to evaluate g_m. The construction shown in Figure 6.15 holds V_{DS} constant at 15 V.

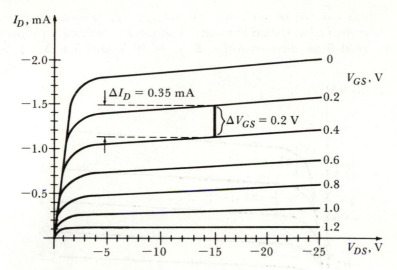

Figure 6.15 Construction for determination of g_m value.

Reading from the curves we find that $\Delta V_{GS} = 0.2$ V. The value of ΔI_D is about 0.35 mA. We can now substitute these numbers into the defining equation for g_m.

$$g_m = \frac{\Delta I_D}{\Delta V_{GS}}\bigg|_{V_{DS}=K} = \frac{(1.55 - 1.2) \text{ mA}}{(0.4 - 0.2) \text{ V}}\bigg|_{V_{DS}=15 \text{ V}}$$

$$= \frac{0.35 \text{ mA}}{0.2 \text{ V}} = 1.75 \text{ mS}$$

We notice that g_m is a ratio of current to voltage and thus has units of siemens (formerly mhos). In this case we have $g_m = 1.750$ mS. The transconductance of a device indicates how much control the gate voltage has over drain current. In a device with high transconductance, the gate is very sensitive in that small gate-voltage changes will cause large drain-current changes.

The dynamic parameter r_d is by definition a ratio of voltage to current. Thus it is called a *resistance*. Since both quantities in the ratio are associated with the drain, we call r_d the drain resistance. The complete name for r_d is the dynamic drain resistance.

The numerical value of r_d is determined graphically from a construction on the characteristic curves. By definition we have

$$r_d = \frac{\Delta V_{DS}}{\Delta I_D}\bigg|_{\Delta V_{GS}=0 \text{ or } V_{GS}=K}$$

By observing the equation for r_d, we can see that our construction will have to be such that the voltage V_{GS} remains constant. The construction shown in Figure 6.16 has the necessary property. We read from the curves that $\Delta V_{DS} = 10$ V and that $\Delta I_D = 50$ µA.

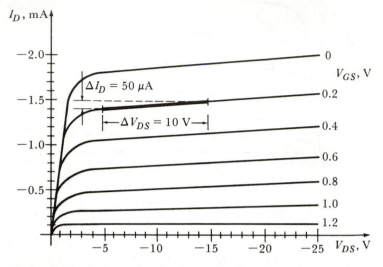

Figure 6.16   Construction for determination of r_d value.

Substituting these values into the defining equation, we obtain the required value of r_d.

$$r_d = \frac{\Delta V_{DS}}{\Delta I_D}\bigg|_{\Delta V_{GS}=0 \text{ or } V_{GS}=K} = \frac{(15-5)\ \text{V}}{(1.5-1.45)\ \text{mA}}\bigg|_{V_{GS}=0.2\ \text{V}}$$

$$= \frac{10\ \text{V}}{50\ \mu\text{A}} = 200\ \text{k}\Omega$$

The parameter named *dynamic drain resistance* is a ratio of V_{DS} to I_{DS}. The quantity is a resistance and in fact is the dynamic resistance seen between drain and source. The quantity r_d can be considered as the dynamic Thévenin resistance as seen looking into the drain and source terminals.

The dynamic parameters are defined in terms of partial derivatives. The delta-quantity equations are exactly true only in the limiting case as delta approaches zero. In the interest of accuracy, the delta quantities should be as small as possible. In our examples, the delta quantities were quite large in order to get better readability from the graphs.

The numerical values of the parameters will depend somewhat on the portion of the characteristic curves used in the calculations. We should use the portion of the characteristic curves near the operating point in order to get accurate values. Although we have not yet talked about choosing an operating point or of making a circuit operate at a chosen point, we will do so in later sections.

6.4 FET BASIC-AMPLIFIER CONNECTION AND THE LOAD LINE

A field-effect transistor can be connected in a circuit that will perform an amplifying function. Generally we think of an amplifier circuit as being one in which the output signal is larger than the input signal. A simple FET amplifier is shown in Figure 6.17. This is a very common configuration. Although we will later refer to this configuration as the common-source connection, we now choose to call this the basic-amplifier connection.

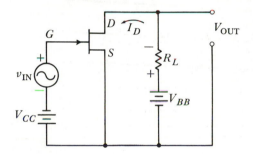

Figure 6.17 Basic-amplifier-circuit connection (N-channel depletion-mode FET).

By inspection of the circuit of Figure 6.17 we see that the output voltage V_{OUT} is taken directly across the drain and source terminals of the FET. Thus we have $V_{OUT} = V_{DS}$. Although most signals to be amplified are ac signals, for the present time we assume that the input voltage and all other quantities are dc quantities. By writing the Kirchhoff-voltage-law (KLV) equation around the loop including the drain and source terminals of the FET and the supply voltage V_{BB}, we obtain

$$V_{DS} + V_{RL} - V_{BB} = 0$$

$$V_{DS} = V_{BB} - V_{RL}$$

$$V_{DS} = V_{BB} - I_D R_L$$

Once the circuit is constructed, the values of V_{BB} and R_L are constant. The equation is therefore seen to be a linear function relating V_{DS} and I_D.

The linear equation relating V_{DS} and I_D for the basic-amplifier-circuit connection is very useful to us. It is so commonly referred to that it is given a special name—the *load-line equation*. The straight-line graph of this equation is called the load line. The load-line equation has the same variables as those normally used on the axes to display FET characteristic curves. It is normal to superimpose the load line on a set of characteristic curves. In Figure 6.18 we have shown a load line drawn on a set of FET characteristic curves.

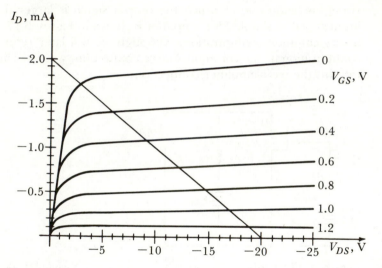

Figure 6.18   FET characteristic curves and load line (*P*-channel depletion-type FET).

We need to formulate some rules so that we can easily and quickly draw a correct load line. Since the load-line equation is always linear and the load line is straight, any two points on the line will define the line. Although any two points will define the load line, we hope to use specific points that are easy to find.

The channel resistance of an FET is controlled by gate voltage. The limiting values of channel resistance are zero and infinity. These limiting values are convenient choices for locating points on a load line. When channel resistance is infinity, the drain current must be zero. When channel resistance is zero, there can be no V_{DS} voltage. The points $I_D = 0$ and $V_{DS} = 0$ are easy to find.

In order to understand the load line, and to become proficient in plotting a load line, an example with known numerical values should be helpful. In Figure 6.19 we have an FET basic-amplifier connection, with numerical values, and a set of characteristic curves. We want to draw the load line for the circuit on the characteristic curves.

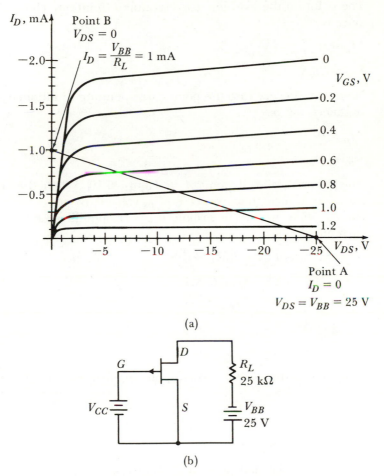

Figure 6.19 Load-line construction: (a) characteristic curves and load line; (b) circuit connection (*P*-channel depletion-type).

Two points on the load line are to be found using limiting values of channel resistance. When the channel resistance is infinity, we know that there can be no drain current. By using the load-line equation, we can solve for V_{DS} when drain current is zero.

$$V_{DS} = V_{BB} - I_D R_L$$
$$= 25 - (I_D)(25 \text{ k}\Omega)$$
$$= 25 - (0)(25 \text{ k}\Omega)$$
$$= 25 \text{ V}$$

The point on the load line corresponding to infinite channel resistance is

$$I_D = 0$$
$$V_{DS} = 25 \text{ V}$$

Note that in every case the point corresponding to infinite channel resistance will be

$$I_D = 0$$
$$V_{DS} = V_{BB}$$

This point is labeled point A in Figure 6.19. When the channel resistance is zero, there can be no V_{DS} voltage. By going to the load-line equation, we can determine the corresponding value of drain current.

$$V_{DS} = V_{BB} - I_D R_L$$
$$V_{DS} = 25 - (I_D)(25 \text{ k}\Omega)$$
$$0 = 25 - (I_D)(25 \text{ k}\Omega)$$
$$(I_D)(25 \text{ k}\Omega) = 25$$
$$I_D = 1 \text{ mA}$$

The point on the load line corresponding to zero channel resistance is

$$V_{DS} = 0$$
$$I_D = 1 \text{ mA}$$

This point is labeled point B in Figure 6.19. Note that the point corresponding to zero channel resistance will always be

$$V_{DS} = 0$$
$$I_D = \frac{V_{BB}}{R_L}$$

The load line is of course a straight line connecting points A and B.
    Any point on the load line defines an operating point for the FET.

For any point on the load line you will find there is a particular set of I_D, V_{DS} values associated with that point. Every possible operating condition for an FET, in a given circuit, is represented somewhere on the load line for that circuit. By looking at the load line for any FET amplifier circuit we can determine the possible range of drain-current values and the possible range of V_{DS} values. The load-line construction gives us quite a lot of information about the operation of an FET amplifier circuit. We will use the load line extensively.

Let us now use the characteristic curves and the load line to develop an intuitive feel or understanding of how amplification occurs with the basic-amplifier circuit. Our discussion will be based on Figure 6.20. A sinusoidal input-signal voltage causes the gate wrt source voltage to leave the Q-point value of 11 V and move along the load line to extreme points A and B. The input-signal voltage waveform is drawn as reflected off the load-line axis and is seen to be of 0.4 V peak-to-peak amplitude. The changing gate wrt source voltage causes the channel resistance of the FET to change, with the result that the channel (drain) current varies in a sinusoidal manner.

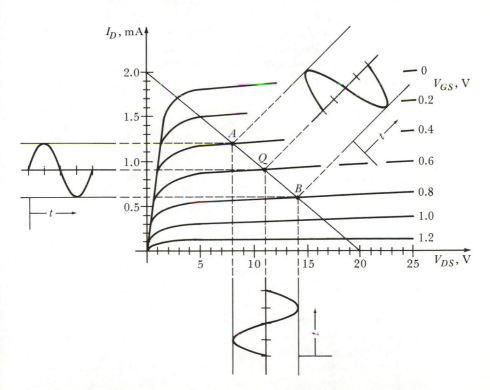

Figure 6.20 A graphical description of the amplification process.

The drain-current waveform is drawn as reflected off the drain-current axis. It is of about 0.6-mA peak-to-peak amplitude.

In the basic-amplifier connection, the drain current must pass through the drain resistor. It will of course produce a voltage drop across the drain resistor. This is important, since we have previously shown that the drain wrt source voltage is equal to the supply voltage V_{BB} minus the voltage drop across the drain resistor. The drain wrt source voltage waveform is drawn as reflected off the V_{DS} axis. The ac portion of this waveform is the output-signal voltage that is seen to be about 6 V peak to peak. With an input signal of 0.4 V peak-to-peak, we have produced an output signal of 6 V peak-to-peak.

Let us sum up the amplification process as explained in the previous two paragraphs. The sinusoidal ac signal voltage applied between gate and source causes the drain current to have a sinusoidal ac component. If the drain resistance is large enough, the drain current produces an output-signal voltage larger than the input-signal voltage.

We have indicated that amplification is dependent on the value of the drain resistance as well as the *I-V* characteristics of the FET. To prove that this is the case, draw a diagram like that of Figure 6.20 but with a load line that corresponds to a small drain resistance.

Hopefully our explanation has been clear and you now understand how a signal voltage can be amplified. We did not attempt to give an explanation of voltage gain in the chapter on BJTs because the explanation requires more steps in the case of a current-controlled device.

6.5   QUIESCENT POINT

The voltage and current values associated with amplifier circuits are many times considered in two distinct and separate parts. These parts are the ac components and the dc components. In many cases the signals to be amplified contain only ac components. The steady-state (that is, dc) voltage and current values are not to be amplified but are necessary in order to make the FET operate properly as an amplifier. An FET, for example, cannot function properly without a dc voltage source to supply drain wrt source voltage. When the circuit is operating properly as an amplifier, the signal (ac) quantities are superimposed on the steady-state (dc) quantities.

When no input signal (ac) is present, the FET and associated circuitry will have only steady-state (dc) voltage and current levels present. Thus the circuit can be considered to be quiet, still, or inactive. These same words are used to describe the meaning of the word *quiescent*. The point on the load line that defines the no-signal I_D, V_{DS} operating point is called the quiescent, or Q, point.

The location of the Q point on the load line is left to the choice of the circuit designer. There are several considerations that may influence the choice of location for the Q point. Among these considerations may be the following:

1. The amplitude of the signal to be handled by the amplifier circuit.

2. The dc power dissipation of the circuit.

3. The linearity of the circuit (that is, the fidelity).

In Figure 6.21 we have a load line drawn on a set of FET characteristic curves. The points on the load line marked Q_1, Q_2, and Q_3 are proposed as being the best choice, respectively, for the three considerations listed previously. We intend to discuss the reasons for the location of points Q_1, Q_2, and Q_3 in separate paragraphs.

Before considering the individual operating points, let us determine what portion of the load line we may legitimately use for our amplifier circuit. In the case we are considering (Figure 6.21), we see that the drain current cannot exceed about 1.85 mA and V_{DS} can never be less than about 2 V. The reason is that the load line does not cross any of the characteristic curves at I_D greater than 1.85 mA or V_{DS} less than 2 V. If the FET device is a junction-gate unit, we will not normally want to let the gate go positive (for N channel, that is). Let us assume that this FET is a junction-gate device and thereby limit our drain current to about 1.75 mA. At the other end of the

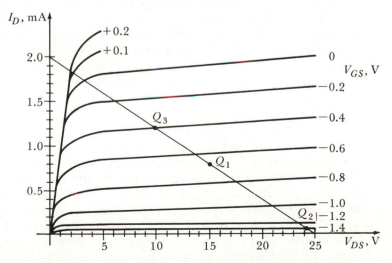

Figure 6.21   *Q*-point locations (*N*-channel depletion-type FET).

load line there is no apparent reason why we cannot use the load line all the way to the intersection with the V_{DS} axis.

In order to keep our amplifier circuit from introducing tremendous distortion, there is a required relationship between the ac signal amplitude and the dc Q-point values. Large distortion is the result of the ac signals attempting to make the instantaneous operating point move off the end of the load line or move into a nonallowed (that is, nonlinear) portion of the characteristic curves. Note that the instantaneous operating point is *not* the Q point but the sum of the steady-state (Q-point) values and the instantaneous ac signal values. The required relationship between ac signal amplitude and Q point is that the sum of the Q-point values and the peak-signal values fall within the allowed region of the load line.

Suppose that we want to design an FET amplifier to handle large signal levels. In this circumstance the point Q_1 in Figure 6.21 is proposed as the best we can do. The ac voltage at the gate wrt source (that is, the peak value of the input-signal voltage) may go as much as 0.7 V in either direction without causing the instantaneous operating point to move into an unallowed region on the load line. This circuit, when biased at Q_1, can handle ac input-signal voltages of 0.7 V peak or 1.4 V peak to peak.

Let us make some observations about the circuit, operating from Q_1, when handling large signals. The corresponding input (gate-source) and output (drain-source) voltages occurring at the instants of the peak of the input waveform are found in the following manner. The peak value of the ac input voltage has been specified as 0.7 V. This makes the total instantaneous gate wrt source voltage to be Q point plus 0.7 V or Q point minus 0.7 V. Numerically we have $V_{GS} = -0.7$ V $+ 0.7$ V $= 0$ V and $V_{GS} = -0.7 - 0.7 = -1.4$ V as the peak instantaneous values of the gate wrt source voltage. Locate these points on the load line. Construct vertical lines from the points just located on the load line to the V_{DS} axis. The intersections will give you the instantaneous peak values of drain wrt source voltage corresponding to the peaks of the input signal. Your results should be $V_{DS} = 24$ V and $V_{DS} = 3$ V.

We have just found that the output voltage swings as much as 9 V above the Q point and as much as 12 V below the Q point. $V_{DS(av)}$, the Q point, is 15 V. The fact that equal input-voltage swings will produce unequal output-voltage swings means that this amplifier circuit is distorting the signal. This particular type of distortion might be very hard to see on an oscilloscope or hear in a speaker. This distortion would be least visible on sinusoidal or square waves. It would be most visible on triangular or sawtooth waveforms. Distortion of this kind is common in circuits handling large signals.

This seems to be a good point to define the ac voltage gain of an amplifier. It is, simply, the magnitude of the ac output voltage divided by the magnitude of the ac input voltage. Any units may be used as long as they are consistent. In this case we have located enough points such that we know that $V_{ds(pp)} = 24 - 3 = 21$ V and $V_{gs(pp)} = 1.4 - 0 = 1.4$ V. Thus the voltage gain A_V is

$$A_V = \frac{V_{out}}{V_{in}} = \frac{V_{ds(pp)}}{V_{gs(pp)}} = \frac{21}{1.4} = 15$$

This tells us that the ac signal at the output of the circuit is 15 times larger in magnitude than the ac signal at the input of the circuit. The same result would have occurred had we used either rms values or peak values.

The quiescent point labeled Q_2 is proposed as being desirable in the case where we want to keep the dc power dissipation as low as possible. The power dissipation can be a most important consideration in some cases—the most obvious being for battery-powered instruments. The dc power dissipated by the FET is the product of I_D and V_{DS}. The power dissipated by the FET is zero at either endpoint of the load line. A few experimental calculations will prove that the power dissipated by the FET is greatest at the center of the load line and goes to zero at the ends. The drain current of our amplifier circuit is also in the load resistance R_L. (See Figure 6.19.) If we choose an operating point near the right end of the load line in Figure 6.21 where the drain current is small, we have simultaneously minimized the power dissipated by the FET and by R_L. Just how close to the end point of the load line Q_2 should be is dependent on the magnitude of the signals to be handled.

Linearity of operation, that is, the fidelity of which the amplifier circuit is capable, can be a valid consideration when choosing a quiescent point. The best linearity and the highest fidelity occur throughout a region where the V_{GS} voltage lines are equally spaced as we move along the load line. In Figure 6.21 the family of curves is such that it is difficult to say just where the lines are most evenly spaced. If the input-signal voltage is limited to 0.2 V peak magnitude, then the quiescent point Q_3 provides quite linear operation.

When choosing a Q point for a FET amplifier circuit, as in many other engineering problems, there are trade-offs involved. For example, if we want a Q point that is optimum for large-signal-handling capability (that is, Q_1), we must trade off some of the efficiency we could obtain at point Q_2 and some of the linearity we could obtain at point Q_3. In general, it is not possible to simultaneously optimize a circuit for each of several conflicting requirements. Usually a compromise point can be found that will be tolerable for every requirement but perhaps not optimum for any single consideration.

6.6   FET BIASING METHODS

As relating to field-effect transistors, bias is defined as being the dc voltage at the gate wrt the source, that is, the dc portion of v_{GS}. By choosing the proper bias voltage, we are able to force the FET and associated circuitry to operate at our chosen Q point. We will consider three different bias methods. The biasing circuit choices we have will depend on whether the FET is an enhancement-mode or depletion-mode device.

FIXED BIAS

The most simple form of bias circuit is the fixed-bias circuit. Fixed bias can be applied equally well to enhancement-mode, depletion-mode, and multimode devices. In the fixed-bias technique we use a dc voltage from a battery, or any other available dc source, and apply it in such a way as to bias the FET properly. In Figure 6.22 we have a

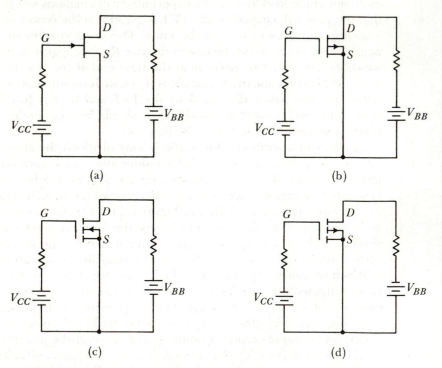

Figure 6.22   Fixed-bias circuit connections: (a) *N*-channel depletion mode; (b) *P*-channel depletion mode; (c) *N*-channel enhancement mode; (d) *P*-channel enhancement mode.

possible configuration with the dc voltage source (V_{CC}) in the gate circuit. In Figure 6.23 the dc voltage source is in the source circuit. The circuits of Figure 6.23 would never be used if the dc bias-voltage source were a battery because the source current would either

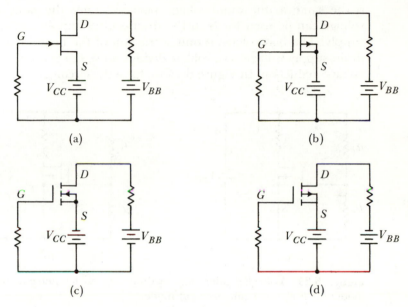

(a)

(b)

(c)

(d)

Figure 6.23 Fixed-bias circuit connections: (a) N-channel depletion-mode JFET; (b) P-channel depletion-mode IGFET; (c) N-channel enhancement-mode IGFET; (d) P-channel enhancement-mode IGFET.

charge or discharge the battery to the point of failure. Regardless of the location of the bias-voltage source, the magnitude of the dc bias voltage needed is dependent on the desired Q-point location. In the special case of an enhancement-depletion device that is to be biased at $V_{GS} = 0$, the circuit of Figure 6.24 is sufficient.

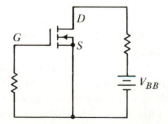

Figure 6.24 Enhancement-depletion N-channel FET biased at $V_{GS} = 0$.

VOLTAGE-DIVIDER BIAS

The voltage-divider biasing method is quite simple. This method is applicable only to MOSFET (IGFET) devices of the enhancement-mode type. For enhancement-mode devices the gate-voltage polarity is the same as the drain-voltage polarity. Thus the same supply voltage can be used for both the drain-source supply and the bias supply. The bias voltage is only a fraction of the magnitude of the drain-supply voltage, so a voltage divider is used to get the necessary voltage reduction. In Figure 6.25 we have the circuit connection for

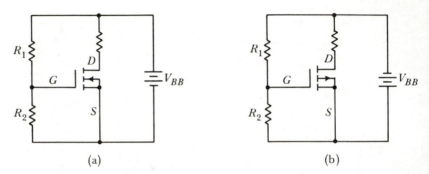

(a)   (b)

Figure 6.25 Voltage-divider bias method: (a) N-channel enhancement mode; (b) P-channel enhancement mode.

voltage-divider bias in both N-channel and P-channel cases. The values of R_1 and R_2 must be chosen properly to obtain the desired Q-point location. By the standard voltage-divider equation, we find that the bias voltage is $V_{GS} = V_{R2} = V_{BB}[R_2/(R_1 + R_2)]$.

SELF-BIAS

The self-bias method is sometimes called the source-bias method. The self-bias name seems appropriate because the source current of a particular FET device is used to develop the voltage necessary for its own biasing. The developed voltage is of such polarity as to bias an FET properly only if the FET is of the depletion-mode type. In case this sounds confusing or impossible, refer to the circuits shown in Figure 6.26. Since there is no voltage drop across R_G (that is, no gate current) we see that the magnitude of V_{GS} is equal to the voltage across R_S. For an N-channel depletion-mode FET we know that V_{GS} should be a negative quantity. This is exactly the case shown in Figure 6.26(a). For a P-channel depletion-mode FET we know that

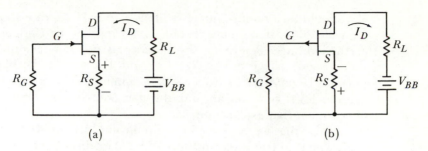

Figure 6.26   Self-bias (source-bias) circuits: (a) N-channel depletion mode; (b) P-channel depletion mode.

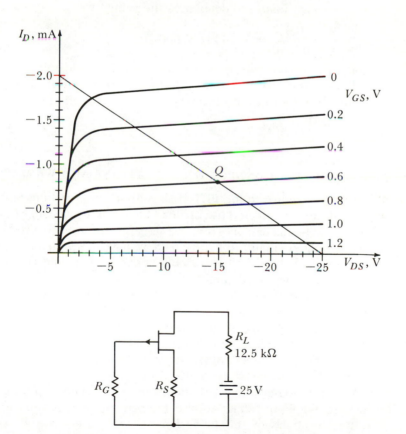

Figure 6.27   Circuit and characteristic curves used to choose R_S value for self-bias at Q.

V_{GS} should be a positive quantity. This is exactly the case shown in Figure 6.26(b). If you have trouble believing that the V_{GS} polarities are as indicated, determine the polarity of V_{SG} remembering that there is no voltage across R_G. V_{GS} and V_{SG} are of the opposite polarity. If you still have trouble determining the polarity, try writing the KVL equation around the gate-source loop or ask your instructor for an explanation.

Suppose that we have an FET circuit like that shown in Figure 6.27. The load line is shown for a 12.5-kΩ load resistance. We want to choose a value for R_S such that the Q point will be as shown on the load line. The chosen Q point requires drain current to be about 0.9 mA, the V_{DS} voltage to be about 14 V, and the V_{GS} voltage to be 0.6 V. The 0.6 V is of course the bias voltage that must be developed across R_S, which has about 0.9 mA of current through it. It is a relatively easy matter to determine the value of R_S.

$$R_S = \frac{V}{I} = \frac{0.6 \text{ V}}{0.9 \text{ mA}} = 0.66 \text{ k}\Omega = 666 \ \Omega$$

This calculation is only approximate, since if we add a 666-Ω source resistance, the load line will be located somewhat differently than shown. In most cases this technique is close enough.

6.7   BASIC-AMPLIFIER CONNECTION: EQUATIONS AND WAVEFORMS

Let us now return to the FET basic-amplifier connection. Since we can now understand and use the load line, we are in a better position to look at the mathematics and the signal waveforms involved in the basic-amplifier circuit. In Figure 6.28 we have the N-channel and P-channel versions of the basic-amplifier connection. The V_{CC} battery polarity is shown for depletion-type devices.

The circuits of Figure 6.28 are obviously two loop circuits. Therefore two equations will be required to completely describe the circuits. When speaking of the input and output voltages, we will be speaking of the voltage of the top terminal wrt the bottom terminal. We will always assume that the input voltage is a pure sinusoidal ac signal with no dc component.

We first choose to consider the input circuit. The loop we choose includes the input-voltage source, the bias supply V_{CC}, and the gate-source terminals of the FET. For the N-channel case in Figure 6.28(a), we obtain

$$-V_{CC} + v_{IN} + v_{SG} = 0 \quad \text{or} \quad v_{GS} = v_{IN} - V_{CC}$$

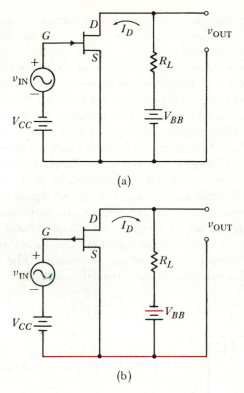

(a)

(b)

Figure 6.28 Basic-FET-amplifier connections: (a) N channel; (b) P channel.

In order to make our equation general enough to include both N-channel and P-channel devices in both depletion and enhancement modes, we need to allow both polarities for V_{CC}. Thus in general

$$v_{GS} = v_{in} \pm V_{CC}$$

Thus we see that the gate wrt source voltage is the sum of a signal voltage and a dc bias voltage. This gate wrt source voltage controls the operation of the FET and thus the entire amplifier circuit.

Using the KVL equation on the output loop of the N-channel FET basic-amplifier connection, we obtain

$$+v_{DS} + i_D R_L - V_{BB} = 0 \quad \text{or} \quad v_{DS} = V_{BB} - i_D R_L$$

The equivalent equation for a P-channel circuit is

$$v_{DS} = -(V_{BB} - i_D R_L)$$

In either case, then, the drain wrt source voltage is always equal in magnitude to the drain-supply voltage less the voltage drop across the load resistor. It will be positive for N-channel devices and negative for P-channel devices.

The signal waveforms associated with the basic FET amplifier are of great interest. These waveforms are, of course, directly related to the equations we have just developed. In most cases we assume that the input signal is sinusoidal in shape. This is a good choice since by Fourier analysis any physically realizable waveform can be expressed as a sum of sinusoids and cosinusoids. In most cases we desire that the output-signal waveform be of the same shape as the input waveform. In many cases we desire the amplitude of the output signal to be greater than the amplitude of the input signal.

In order to improve our understanding of the basic-amplifier circuit, let us look at an actual circuit with all numerical values known. We will apply a known input-signal voltage waveform and then determine the output-signal voltage waveform. The circuit shown in Figure 6.29 is the circuit under consideration. The charac-

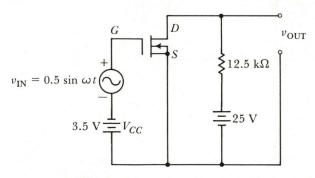

Figure 6.29   FET basic-amplifier circuit (N-channel IGFET).

teristic curves and load line are shown in Figure 6.30. The 1-V peak-to-peak input signal causes the instantaneous operating point to move between the extreme points A and B. The points A, B, and Q can of course be translated into V_{DS}, V_{GS}, and I_D values. Since we know the waveshape of the input-signal voltage, we can translate the information contained in Figure 6.30 into the waveforms shown in Figure 6.31.

Let us make some observations with regard to the waveforms of Figure 6.31. We note that the input-signal voltage v_{IN} differs from v_{GS} by a constant amount of 3.5 V. This of course agrees with the equation derived previously for the input loop. The drain current is

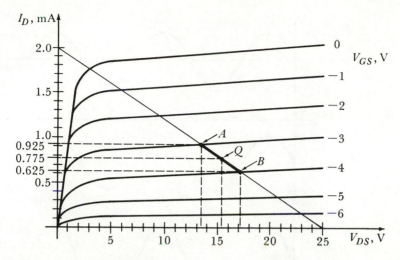

Figure 6.30 Characteristic curves and Q point for the circuit of Figure 6.29.

in phase with the input-signal voltage because as the input goes positive, the instantaneous operating point moves from Q toward point A increasing drain current. The v_{DS} voltage is seen to be 180° out of phase wrt the input-signal voltage. As the input-signal voltage goes in a positive direction, the instantaneous operating point moves toward point A. Moving from point Q to point A constitutes a decrease in v_{DS}. Thus we see that as the input-signal voltage moves in a positive direction, v_{DS} moves in a negative direction.

The ac voltage gain of an amplifier is defined as the ac output-signal voltage divided by the ac input-signal voltage. In this particular case we will choose to use peak-to-peak units to find that the voltage gain is

$$A_V = \frac{V_{\text{out}}}{V_{\text{in}}} = \frac{3.5\ V_{\text{pp}}}{1\ V_{\text{pp}}} = 3.5$$

With the exception of the input-signal voltage waveform, all the waveforms of Figure 6.31 have a dc level. The dc levels are not usually desirable but are necessary. We must have some dc drain wrt source voltage so that we can have some drain current for v_{GS} to control. Note that we did not consider any dc levels when we calculated the gain in the previous paragraph. This is so because we are interested (usually) in amplifying only the ac signals.

Let us consider the waveforms in terms of their mathematical

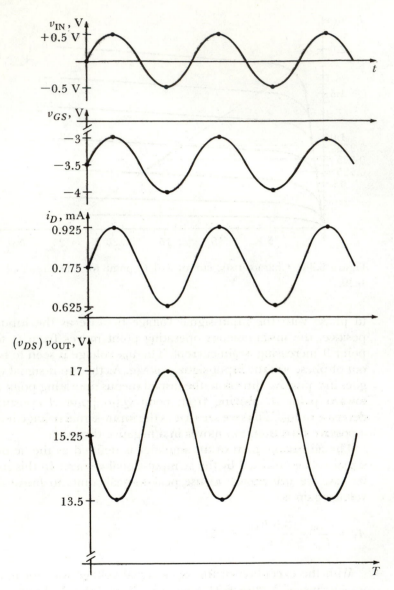

Figure 6.31   Waveforms for the circuit of Figure 6.29 as obtained from the characteristic curves of Figure 6.30.

description. We have determined that the output signal is 3.5 times larger than the input signal. We also note that the output has an added 15.25-V dc level and the output signal is 180° out of phase wrt the input signal. The input- and output-signal voltages are related as follows.

$$v_{OUT} = -3.5v_{IN} + 15.25$$

In this case

$$v_{IN} = 0.5 \sin \omega t$$

We can then express the output voltage as

$$v_{OUT} = -3.5(0.5 \sin \omega t) + 15.25$$
$$= -1.75 \sin \omega t + 15.25$$

or as

$$v_{OUT} = 1.75 \sin (\omega t + 180°) + 15.25$$

The equation with the minus sign, used to convey the polarity information, is more compact and in many cases preferable. The drain current is

$$i_D = (0.15 \sin \omega t + 0.775) \text{ mA}$$

As a check on the validity of our work, we will solve the load-line equation in terms of the waveform quantities obtained from the characteristic curves. The load-line equation for an N-channel device is

$$V_{DS} = V_{BB} - I_D R_L$$

Substituting the mathematical descriptions for V_{BB} and for i_D, we obtain

$$v_{DS} = 25 - [(0.15 \sin \omega t + 0.775) \text{ mA}] \times (12.5 \text{ k}\Omega)$$
$$= 25 - 1.875 \sin \omega t - 9.69$$
$$= -1.875 \sin \omega t + 15.31 \text{ V}$$

This equation is probably not exactly what we would obtain directly from the load line and characteristic curves, or from Figure 6.31, but it is close. The errors are of course a function of the accuracy with which the graphs are constructed and read. In the equation for v_{DS}, remember that the minus sign represents polarity information, the 1.875 is a peak value, and the 15.31 represents a dc level.

6.8   AC EQUIVALENT CIRCUITS

In many cases where amplifier circuits are needed, we are interested in amplifying only ac signals. Under these conditions it would be highly undesirable for the circuit to amplify or introduce any dc levels that happened to be present. The introduction of capacitors at the input and output of our basic-amplifier-circuit connection will make certain that no dc levels enter or leave the amplifier circuit. These capacitors allow us to use any of the biasing techniques we have previously discussed. In Figure 6.32 we have an FET basic-amplifier circuit with dc levels blocked by capacitors.

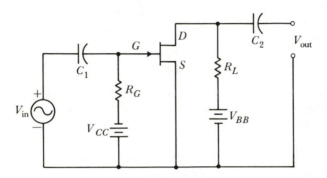

Figure 6.32   ac-coupled FET amplifier circuit (N-channel depletion-type FET).

Consider the function of the capacitors in the circuit of Figure 6.32. On the input side we note that the capacitor C_1 will block any dc level that might be present at the signal source from being seen or felt by the amplifier circuit. The capacitor C_1 is also effective in keeping any bias voltage from being disturbed or shorted by the signal source. The capacitor C_2 performs a similar function at the signal-output end of the amplifier circuit. The capacitors used in this application are called *coupling capacitors*. The capacitance of the capacitors is chosen so that they are essentially short circuits (that is, X_C is very small) at the signal frequency of interest.

We have seen that coupling capacitors prevent any dc levels from appearing at the output terminals. Since this is the case, we note that the mathematical relationship between input and output voltages can be simplified to $V_{out} = A_v V_{in}$. This equation applies only to the case where the amplifier is ac-coupled and where the voltages V_{out} and V_{in} are ac voltages.

The schematic drawing can be simplified when we are considering only ac signals. The simplification results from the fact that the coupling capacitors and the batteries are short circuits at the ac signal frequencies. The simplified circuit diagram is called an ac equivalent circuit since it applies only to the case where ac signals are being considered. In Figure 6.33 we have the ac equivalent of the circuit shown in Figure 6.32.

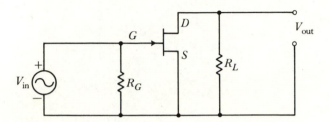

Figure 6.33 ac equivalent of basic-amplifier circuit.

Our ac equivalent circuit will be complete when we are able to find an ac equivalent for the FET. In order to find an ac equivalent representation, let us look once again at the dynamic parameters that define the small-signal ac operation of FET devices.

$$\mu = \frac{\Delta V_{DS}}{\Delta V_{GS}}\bigg|_{\Delta I_D = 0}$$

$$g_m = \frac{\Delta I_D}{\Delta V_{GS}}\bigg|_{\Delta V_{DS} = 0}$$

$$r_d = \frac{\Delta V_{DS}}{\Delta I_D}\bigg|_{\Delta V_{GS} = 0}$$

Note that the quantity ΔI_D appears in the equations for both g_m and for r_d. Since we are interested in an ac equivalent, and since the delta quantities are essentially ac quantities, let us solve for the quantity ΔI_D.

$$\Delta I_D = g_m \Delta V_{GS}$$

and

$$\Delta I_D = \frac{\Delta V_{GS}}{r_d}$$

The equations seem to indicate that the drain current is in two distinct parts. One part of the drain current is a function of gate wrt source voltage and the other part of the drain current is seen to be a function of drain wrt source voltage. In Figure 6.34 we have an ac

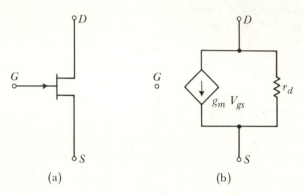

(a)   (b)

Figure 6.34 FET representation: (a) normal FET symbol; (b) ac equivalent of FET.

equivalent circuit of an FET based upon the two equations describing drain current I_D. We will not be using the Δ (delta) symbol with our ac equivalent because we have already restricted the use of the circuit to the ac case, and ac signals are certainly changing quantities, as is implied by the Δ symbol. Note that the ac equivalent we have chosen to represent the FET does in fact show the drain current in two distinct parts. Each part is as indicated by solving the defining parameter equations for I_D.

The ac equivalent circuit for an FET, as shown in Figure 6.34(b), applies to N-channel and P-channel devices as well as to both depletion-mode and enhancement-mode operation. This is the current source or the Norton ac equivalent circuit of an FET. Note that the gate terminal is open-circuited. The gate current is zero.

Another ac equivalent circuit of an FET is commonly used. It is the voltage-source type or the Thévenin equivalent of the FET. The voltage-source version is obtained from the current-source version by performing a conversion from a Norton equivalent to a Thévenin equivalent. In Figure 6.35 we have two alternate forms of the voltage-source equivalent circuit. When the Norton-to-Thévenin conversion is made, you obtain a voltage source of magnitude $g_m r_d V_{gs}$. Since it is true that $\mu = g_m r_d$, the results shown in Figure 6.35 are correct. The polarity of the voltage source may be handled by either of the methods shown in Figure 6.35.

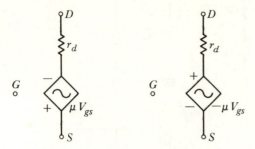

Figure 6.35   Voltage-source ac equivalent circuits for an FET.

EXAMPLE 6.1   A P-channel MOS depletion-mode FET is to be used in a basic-amplifier-circuit configuration. The various circuit quantities are as follows:

$\mu = 50$

$g_m = 1000 \ \mu S$

$R_L = 2 \ k\Omega$

$R_G = 100 \ k\Omega$

Using first the voltage-source equivalent circuit for the FET and then the current-source equivalent, find the ac voltage gain of the amplifier circuit.

SOLUTION   As a first step we will draw the schematic diagram of the amplifier circuit, as shown in Figure 6.36(a). Since we are interested only in the ac operation, let us draw the ac equivalent of the circuit except for the FET. This is shown in Figure 6.36(b).

At this point let us divide the problem into parts I and II as we use the various FET equivalents.

PART I   We will draw the entire ac equivalent of the circuit using the voltage-source equivalent for the FET. This has been done in Figure 6.36(c). Note that we will need a numerical value for r_d. We can calculate r_d since we know that

$\mu = g_m r_d$

$$r_d = \frac{\mu}{g_m} = \frac{50}{1000 \times 10^{-6}} = 50 \times 10^3 \ \Omega$$

Note that for this connection the voltage V_{in} is the same as V_{gs}.

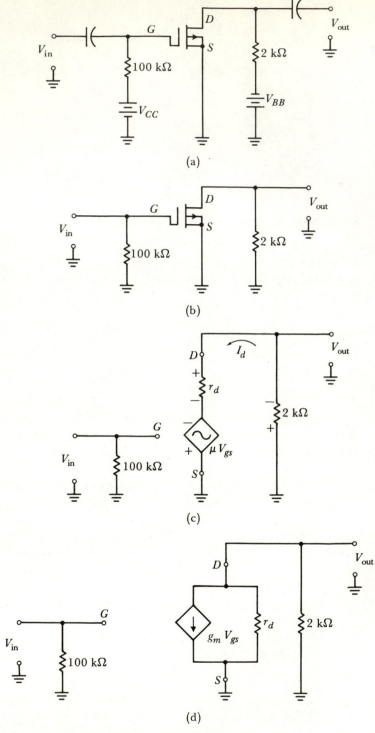

Figure 6.36 Circuit for Example 6.1.

Around the loop including the drain, the source, and R_L we have one voltage source and two resistances. The ac drain current is in the direction shown. The magnitude of drain current is

$$I_d = \frac{\mu V_{gs}}{r_d + R_L} = \frac{50 V_{in}}{50,000 + 2000}$$

$$= \frac{50 V_{in}}{52 \times 10^3} = 0.962 \times 10^{-3} V_{in}$$

The output voltage is taken across R_L so that the output ac voltage is

$$V_{out} = I_d R_L$$

$$= 0.962 \times 10^{-3} \times 2 \times 10^3$$

$$= 1.924 V_{in}$$

The voltage gain is defined as output voltage divided by input voltage. Thus the ac voltage gain is

$$A_V = \frac{1.924 V_{in}}{V_{in}} = 1.924$$

In case there has been any confusion in your mind about the difference between amplification factor μ and the amplification of a circuit, this example should help clarify the situation. In this case the amplification factor of the FET is 50 whereas the amplification (gain) of the amplifier circuit is 1.924.

PART II Part II of the solution differs from part I only in the choice of FET equivalent. We now draw the complete ac equivalent of the amplifier circuit using the current-source equivalent for the FET as shown in Figure 6.36(d).

In solving this problem we need to know V_{gs}. Observation of the ac equivalent circuit shows that $V_{in} = V_{gs}$. In the output portion of this circuit we notice that we have a current source $g_m V_{gs}$ in parallel with two resistances. We can solve for the ac current through the load resistance by applying the current division rule.

$$I_{rl} = g_m V_{gs} \frac{50 \text{ k}\Omega}{(50 + 2)\text{k}\Omega}$$

$$= 1000 \ (10^{-6}) \ V_{in} \frac{50 \text{ k}\Omega}{52 \text{ k}\Omega}$$

$$= 962 \ (10^{-6}) \ V_{in}$$

The output voltage is taken across R_L so that we have

$$V_{out} = I_{rl}R_L$$
$$= 962 \times 10^{-6} \times 2 \times 10^3\ V_{in}$$
$$= 1.924\ V_{in}$$

The ac voltage gain is

$$A_V = \frac{V_{out}}{V_{in}} = \frac{1.924\ V_{in}}{V_{in}}$$
$$= 1.924$$

The results of the calculations of parts I and II are the same. This is to be expected, since the only change involved a Thévenin-to-Norton conversion in the FET equivalent.

EXERCISES

QUESTIONS

Q6.1 What differences should a user note between N-channel and P-channel FETs?

Q6.2 What are the constructional differences between JFETs and IGFETs?

Q6.3 What is the channel in an FET?

Q6.4 What is meant by the term *pinch off* as related to FETs? Your answer is to include both gate-source and drain-source types of pinch off.

Q6.5 Since an enhancement-mode IGFET has no significant channel as manufactured, how is it possible to get a drain-source current?

Q6.6 What is meant by a normally-on type of MOSFET as compared with a normally-off type of MOSFET?

Q6.7 In what way do the schematic symbols for IGFETs show whether the represented device is an enhancement-mode or a depletion-mode one?

Q6.8 How do supply-voltage polarities differ for N-channel and P-channel FETs?

Q6.9 How do required supply voltages vary for enhancement-mode FETs as compared with depletion-mode FETs?

Q6.10 Why is the name *amplification factor* appropriate for the FET dynamic parameter whose letter symbol is μ?

Q6.11 Why is the name *transconductance* appropriate for the FET dynamic factor whose letter symbol is g_m?

Q6.12 Why is the name *dynamic drain resistance* appropriate for the FET parameter whose letter symbol is r_d?

Q6.13 When determining dynamic parameter values from a set of characteristic curves, why should we keep the delta quantities as small as convenient?

Q6.14 Based on the load-line equation and/or a load-line plot, how can you tell that an FET basic-amplifier circuit inverts the signal-voltage polarity?

Q6.15 Why is it that the ac load line must go through the Q point on the dc load line?

Q6.16 What are some factors to consider when choosing a Q point for an FET basic-amplifier circuit?

Q6.17 What bias methods are suitable for depletion-mode FETs?

Q6.18 What bias methods are suitable for enhancement-mode FETs?

Q6.19 Which of the listed bias methods have a self-adjusting feature?

Q6.20 Why is there no connection to the gate terminal of the ac equivalent circuit of an FET?

PROBLEMS

P6.1 Draw the schematic symbols for both N-channel and P-channel JFET devices. Show the voltage polarities needed and the current directions involved.

P6.2 Draw the schematic symbols for both N-channel and P-channel enhancement-mode IGFET devices. Show voltage polarities and current directions. Assume enhancement-mode devices.

P6.3 Repeat Problem 6.2 for depletion-mode IGFETs.

P6.4 Use graphical techniques to find μ for an N-channel JFET whose characteristics are shown in Figure 6.37. Assume that the FET has about 1 mA of current and about 15 V across it.

P6.5 Find g_m for a P-channel enhancement-mode IGFET whose curves are shown in Figure 6.38. Assume that the device is operating with a Q point of $V_DS = -20$ V, $I_D = -0.5$ mA.

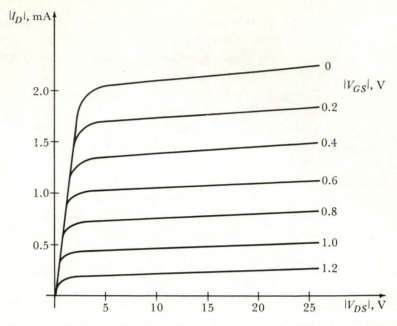

Figure 6.37   Depletion-mode FET characteristics for Chapter 6 problems.

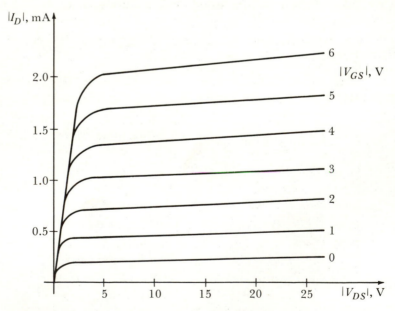

Figure 6.38   Enhancement-mode FET characteristics for Chapter 6 problems.

P6.6 Find r_d for the N-channel multimode IGFET whose characteristics are shown in Figure 6.39.

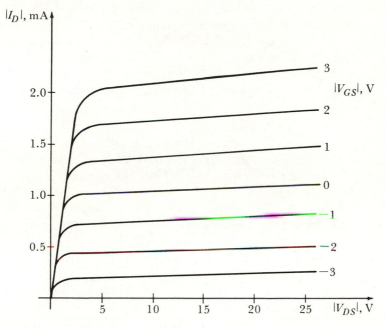

Figure 6.39   Multimode FET characteristics for Chapter 6 problems.

P6.7 Use Kirchhoff's voltage law to derive the load-line equation for an N-channel enhancement-mode IGFET connected in a basic-amplifier configuration. The FET is characterized in Figure 6.38.

P6.8 Derive the load-line equation for a P-channel depletion-mode JFET connected in a basic-amplifier configuration. The FET is as described in Figure 6.37.

P6.9 Plot the load line for an N-channel depletion-mode IGFET in a basic-amplifier circuit. The FET is characterized in Figure 6.37. Assume $V_{BB} = 15$ V and $R_L = 5$ kΩ.

P6.10 Plot the load line for a P-channel depletion-mode JFET in a basic-amplifier circuit. $V_{BB} = 25$ V, $R_L = 25$ kΩ. Characteristic curves as shown in Figure 6.37.

P6.11 Plot the load line for an N-channel enhancement-mode MOSFET when $V_{BB} = 10$ V and $R_L = 20$ kΩ. Use curves of Figure 6.38.

P6.12 A basic-amplifier circuit is to be biased with a fixed-bias technique so that $V_{DS} = 10$ V. The FET is a P-channel JFET with characteristics shown in Figure 6.37. The supply voltage V_{BB} is 25 V. The drain resistance R_D is 10 kΩ. Draw the schematic of the complete circuit you would use including all numerical values.

P6.13 An N-channel enhancement-mode IGFET is to be biased so that the Q point is at $I_D = 1$ mA. The FET curves are shown in Figure 6.38. Use the voltage-divider bias method. The supply voltage $V_{BB} = 20$ V and the drain resistance $R_D = 10$ kΩ. Calculate all component values needed and draw the schematic diagram of the circuit.

P6.14 An FET basic-amplifier circuit is to have a source resistance for increased bias stability. The FET is an N-channel enhancement-mode MOSFET with curves like those of Figure 6.38. The circuit is to be biased so that $V_{DS} = 15$ V. Other factors are $V_{BB} = 25$ V, $R_D = 10$ kΩ, $R_S = 2.500$ kΩ. Use the voltage-divider bias technique. Calculate all component values needed and draw the schematic diagram of your proposed circuit.

P6.15 A P-channel multimode MOSFET is connected in a basic-amplifier configuration. It is to be biased so that V_{DS} is near 15 V. The FET specifications are as shown in Figure 6.39. Other values are $V_{BB} = 24$ V, $R_D = 8$ kΩ. Use any bias method you choose. Draw the circuit diagram you propose showing all numerical values.

P6.16 An N-channel depletion-mode JFET is connected in a basic-amplifier-circuit configuration. $V_{BB} = 20$ V. It is to be biased so that at the Q point, $V_{GS} = 0.8$ V. The load line is to intersect the I_D axis at 1.5 mA. Use the self-bias technique. Draw the complete circuit showing all numerical values, including R_D.

P6.17 Graphically determine the voltage gain for the circuit of Problem 6.13 where $V_{out} = V_{DS}$.

P6.18 Determine the voltage gain of the circuit of Problem 6.14. The output voltage is taken at the drain wrt ground, not V_{DS}.

P6.19 Repeat Problem 6.18 for zero source resistance.

P6.20 Find the voltage gain of the circuit of Figure 6.40 using the Thévenin ac equivalent circuit. The FET parameters are $g_m = 2000$ μS and $r_d = 100$ kΩ.

P6.21 Use ac equivalent circuit methods to find the voltage gain of the circuit of Figure 6.41. The FET is described as $g_m = 5$ mS, $\mu = 500$, and $r_d = 100$ kΩ.

P6.22 Repeat Problem 6.21 to find current gain for a 50-kΩ load resistor connected between the output terminals. Input current is

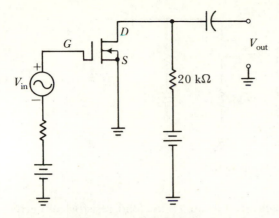

Figure 6.40   See Problem 6.20.

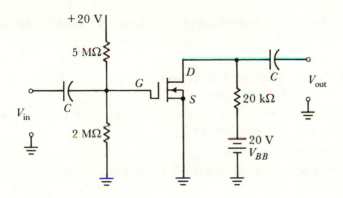

Figure 6.41   See Problem 6.21.

defined as the current delivered to the signal-input terminals and output current is the current through the 50-kΩ load resistor.

P6.23 Repeat Problem 6.21 except find power gain. Input power is defined as the signal power delivered to the signal-input terminals and output power is the signal power delivered to a 10-kΩ load resistor connected across the output terminals.

P6.24 Find the voltage gain of the circuit of Figure 6.42 by ac equivalent circuit methods. The parameters describing the FET are $\mu = 500$ and $g_m = 2.5$ mS.

P6.25 Find the current gain of the circuit of Figure 6.29 if the FET parameters are $\mu = 400$ and $g_m = 2$ mS. The output current is the

256   FIELD-EFFECT TRANSISTORS

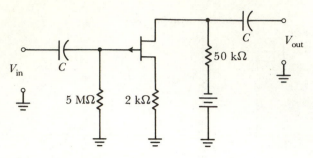

Figure 6.42   See Problem 6.24.

signal current delivered to an 18 kΩ load resistor connected across
the signal-output terminals.

P6.26 Calculate the power gain for the situation described in Problem 6.25.

P6.27 Use ac-equivalent-circuit methods to calculate the output
impedance of the circuit of Figure 6.41 if

$\mu = 200$ and $r_d = 20$ kΩ.

P6.28 Repeat Problem 6.27 to find the input impedance. Is input Z
a function of FET parameters?

P6.29 Find the output impedance for the circuit of Figure 6.42 if
the FET has characteristics of

$\mu = 100$ and $g_m = 1.5$ mS

P6.30 Repeat Problem 6.29 to find the input impedance.

CHAPTER 7
VACUUM TRIODES

The vacuum triode is very closely related to the vacuum diode. The triode is basically a diode with an added element, the control grid. Even though the triode differs from the diode in a small way, the invention of the triode is considered to be one of the great inventions of all time.

Lee DeForest added a third element, called the grid, to a vacuum diode in 1905 or 1906. The resulting component was called an Audion or triode. The word *triode* refers to a three-element vacuum tube in the same way that the word *diode* refers to a two-element vacuum tube. The triode is without doubt the basic invention upon which the whole electronics field is built. The triode made possible amplification of electric signals. Certainly the importance of amplification to the electronics field cannot be overemphasized.

Vacuum-triode usage has decreased greatly as BJT- and FET-type devices have been developed. This trend will probably continue. Most newly designed electronic circuits will contain BJTs and/or FETs, either as individual components or as part of integrated circuitry, rather than vacuum tubes.

Such circumstances have led many authors to delete from their textbooks any study of vacuum triodes. The inclusion of a discussion of vacuum triodes is no doubt considered unnecessary and the deletion of such indicates that the book is modern. Obviously we have not taken this approach.

There are untold millions of vacuum tubes in use every day in many different applications. Technology graduates for a good many years to come should have reasonable expectation of being called upon to apply their technical skills to complex equipment containing vacuum tubes. This being the case, we have given the vacuum triode reasonable coverage. When the vacuum triode topics of this chapter are essentially the same as the analogous topic in the FET chapter, we will refer the reader to the FET chapter. This will save time and space and still allow the student to become thoroughly familiar with vacuum triodes. As we will see later, a vacuum triode is quite analogous to an N-channel depletion-mode FET.

7.1   VACUUM-TRIODE CONSTRUCTION

The construction of a vacuum triode is similar to that of a vacuum diode. As in the case of the vacuum diode, the triode can be built with either a directly heated or an indirectly heated cathode. Directly heated cathodes are usually used in portable-battery-operated equipment. In order to minimize the filament power consumption and thus the size of the filament battery needed, directly heated cathodes (that is, the filaments) are rather small. They tend to be fragile. Indirectly heated cathodes are used in equipment operated from ac line voltage. They are more rugged and the power consumption is not usually a critical factor. If ac were used as filament power in a directly heated cathode tube, the ac filament voltage would be interpreted as a signal to be amplified. In a radio this would be noticed as a hum at the speaker.

A typical vacuum tube triode is shown in Figure 7.1. The elements

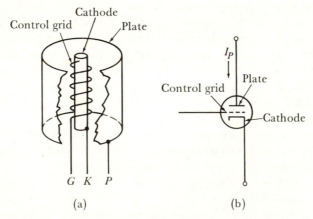

Figure 7.1   Vacuum triode (a) construction and (b) schematic symbol.

are placed in an evacuated envelope. The elements shown must be supported mechanically in the proper position. Some of the physical supporting devices are made of mica (or some other material), which is an electrical insulator and is capable of surviving the high temperatures involved.

The cathode and the plate in a triode tube perform the same functions as in a diode tube. Thus the heated cathode boils off the electrons that are attracted by the positive voltage normally on the plate. Negative charges (electrons) move from cathode to plate. We will normally say that conventional current is from plate to cathode.

The control grid is many times a spiral of wire as shown in Figure 7.1. It is physically placed between the cathode and plate. By changing the voltage between grid and cathode, we find that the control grid can control the amount of current through the tube. As mentioned before, this control concept is very important.

Not all tubes are constructed as in Figure 7.1. Some have a construction so that the parts are physically placed something like the schematic symbol of a tube shown in Figure 7.1(b). These are called *planar tubes*. Figure 7.2 shows a wide variety of vacuum-tube sizes and construction methods.

7.2 VACUUM-TRIODE *I-V* CHARACTERISTICS

There are three quantities that are interrelated to form the *I-V* characteristics of vacuum tubes: the plate current I_P, the plate wrt cathode voltage V_{PK}, and the grid wrt cathode voltage V_{GK}. These three variables will be displayed on a two-dimensional graph in a manner similar to that already done for BJTs and FETs. The running variable will be the grid wrt cathode voltage V_{GK}.

In Figure 7.3 we have the *I-V* characteristics of a 12AX7 vacuum triode. This type of information is normally supplied by the manufacturer but can be experimentally determined if desired. In the case of a 12AX7, there are two identical devices within the same vacuum envelope. As indicated in the figure, the *I-V* characteristics apply to each of the sections.

Figure 7.3 indicates that V_{GK} is opposite in polarity as compared with V_{PK}. This is true for all normal vacuum triodes and indicates that the device is of the normally-on type. By analogy with FETs, then, we must conclude that vacuum triodes are available only as depletion-mode devices.

As indicated by Figure 7.3, a vacuum triode will conduct quite heavily when $V_{GK} = 0$. As V_{GK} takes on negative values, the negative

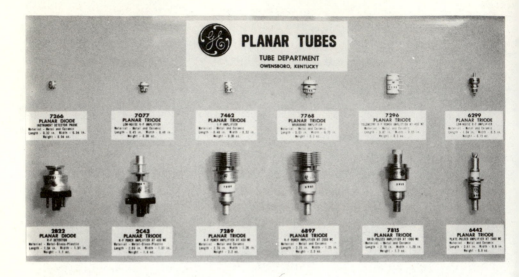

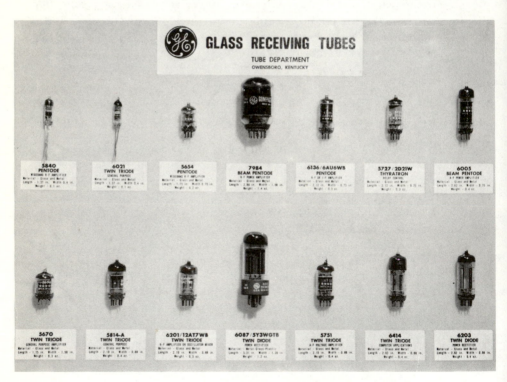

Figure 7.2 Vacuum tubes of various sizes and construction methods. (Courtesy of General Electric Company.)

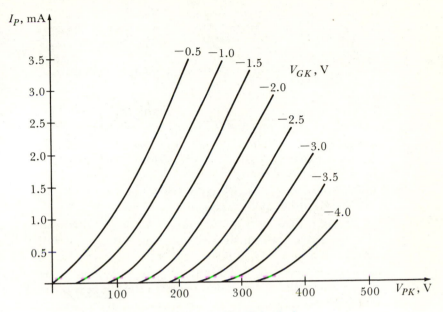

Figure 7.3 *I-V* characteristics of 12AX7-A triode.

voltage tends to repel electrons back toward the cathode (which emits them), thereby diminishing the plate current. If V_{GK} takes on large-enough negative values, the plate current will cease and the tube is said to be "in cutoff." This is analogous to the gate-source pinch-off voltage in FETs. According to the curves, a 12AX7 with 350 V of V_{PK} will be in cutoff when V_{GK} is negative by 4.5 (or more) V.

7.3   VACUUM-TUBE PARAMETERS

There are three variables of great interest to us in the study of vacuum tubes. They are the voltage at the plate wrt cathode, the voltage at the grid wrt cathode, and the plate current. The voltage variables will be double subscripted to indicate what voltage is being referred to. The definition of our method of double-subscripting and several examples are given in Chapter 0.

We define the vacuum-tube parameters in terms of the quantities I_P, V_{PK}, and V_{GK}. The parameters are defined as follows:

$$\mu = \frac{\Delta V_{PK}}{\Delta V_{GK}}\bigg|_{\Delta I_P = 0} = \frac{\partial V_{PK}}{\partial V_{GK}} = \text{amplification factor}$$

$$g_m = \frac{\Delta V_P}{\Delta V_{GK}}\bigg|_{\Delta V_{PK}=0} = \frac{\partial V_P}{\partial V_{GK}} = \text{transconductance}$$

$$r_p = \frac{\Delta V_{PK}}{\Delta I_P}\bigg|_{\Delta V_{GK}=0} = \frac{\partial V_{PK}}{\partial I_{PK}} = \text{dynamic plate resistance}$$

The parameters μ, g_m, and r_p are dynamic quantities. Thus they are defined in terms of changes of quantities. Strictly speaking, the partial-derivative definition is the exact definition. The delta-(change-of) quantity definition is very close for small delta quantities, however, and has the advantage of being numerically evaluated rather easily from a set of characteristic curves.

The vacuum triode is analogous to an FET. The plate, cathode, and grid of a vacuum triode are analogous to the drain, source, and gate, respectively, of an FET. This being the case, it should be no surprise that the dynamic parameters for vacuum triodes are defined like those for FETs. If you have forgotten the physical implications of each parameter, refer to the chapter on FETs.

Numerical values can be found for each dynamic parameter by a construction on the I-V characteristic curves of a triode. The required construction is quite similar to that undertaken for FETs.

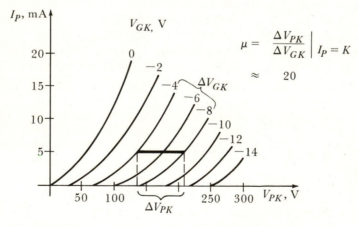

Figure 7.4 Graphical determination of μ.

Since the characteristic curves are somewhat different for vacuum triodes, we have chosen to make the constructions necessary to find μ, g_m, and r_p. These are shown in Figures 7.4 to 7.6.

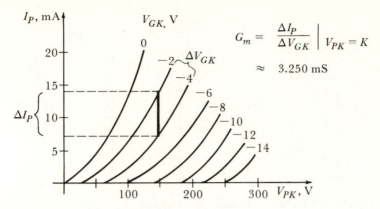

Figure 7.5 Graphical determination of transconductance g_m.

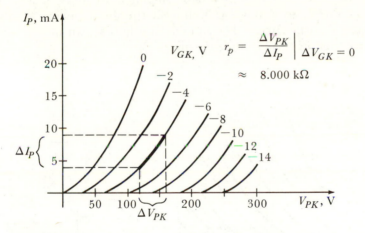

Figure 7.6 Graphical determination of dynamic plate resistance r_p.

7.4 BASIC-AMPLIFIER CONNECTION AND THE LOAD LINE

As we already know, vacuum tubes can be connected to become part of an amplifying circuit. The circuit of Figure 7.7 shows a vacuum tube in a commonly used amplifying circuit that we call the basic-amplifier connection. The input-signal voltage is applied between grid and cathode. A dc level is also applied between grid and cathode to provide the proper dc operating (bias) levels. The output-signal voltage is the ac portion of the voltage waveform at the plate wrt ground.

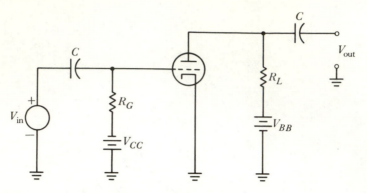

Figure 7.7   Vacuum-tube-amplifier circuit.

In Figure 7.8 we have a basic-amplifier circuit. Numerical values are shown so that we can draw a specific load line on the *I-V* characteristic curves. The numerical values shown may be typically expected in vacuum-tube circuits.

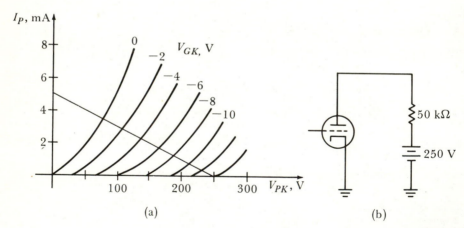

Figure 7.8   Vacuum-tube-amplifier circuit: (a) characteristics and load line; (b) circuit connection.

Since the vacuum triode is analogous to the FET, the basic-amplifier connection using vacuum triodes is analogous to the basic-amplifier connection we studied in the FET chapter. Therefore we will not take the time and space to write the load-line equation and plot the load line as we have previously done. Please refer to the FET chapter if you need such information.

You may wonder why we keep referring to the vacuum triode as being analogous to an FET instead of a BJT. A look at the characteristic curves will indicate the reason. In all three cases, the *I-V* characteristics show the current in the device as the ordinate. Also in each case the terminal voltage across the device is plotted as the abscissa. A distinct difference will be noted when comparing the running variable in the three cases. For the FET and the vacuum tube, the running variable is a voltage. For the BJT, the running variable is a current.

The vacuum triode and the FET are seen to be analogous because they are voltage-controlled devices. Their respective control elements (grid and gate) draw no current. The vacuum tube and the *N*-channel depletion-mode FET are directly analogous. Some manufacturers even make FET devices that are packaged to plug directly into the socket of older vacuum-tube equipment. Vacuum tubes and FETs are not usually interchangeable in existing equipment, however, because of packaging differences and because FETs do not usually have the high-voltage capability of vacuum tubes.

7.5 QUIESCENT POINT

The quiescent, or *Q*, point for vacuum-triode circuits has a meaning similar to that for BJTs and FETs. Three common criteria you might want to consider in choosing a *Q* point are

1. Signal amplitude-handling ability.
2. Linearity (fidelity).
3. Power dissipation.

Since we have previously discussed each of these criteria, we will not repeat the discussion here.

7.6 VACUUM-TRIODE BIASING METHODS

As relating to vacuum tubes, bias is defined as the dc voltage at the grid wrt the cathode. It is assumed that the grid is to be negative wrt the cathode, since tubes are analogous to *N*-channel depletion-mode FETs. By choosing the proper bias voltage, we are able to force the vacuum tube and the circuit to operate at our chosen *Q* point. Any bias method that applies to the depletion-type FET will apply to vacuum triodes. In addition, there are some distinct vacuum-triode biasing circuits.

FIXED BIAS

Fixed-bias circuits for vacuum triodes are the same as for FETs. See the FET chapter for details.

GRID-LEAK BIAS

At first glance, a circuit meant to operate with grid-leak bias appears to have no bias provisions at all. Such a circuit is shown in Figure 7.9. V_{GK} is sinusoidal, with zero average or dc level, as shown in Figure 7.9(b), when the tube is removed from its socket. If the tube is installed in the socket, the grid will draw current when it is positive

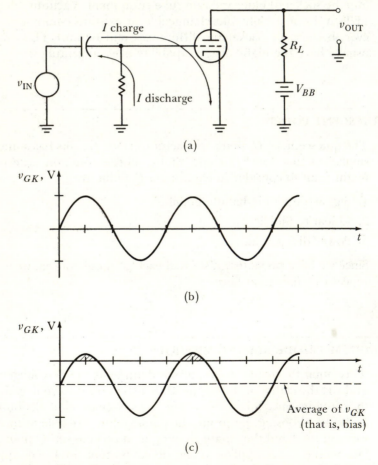

Figure 7.9   Grid-leak-bias circuit. (a) Circuit diagram; (b) voltage between grid and cathode pins of socket with tube removed from socket; (c) V_{GK} in normal operation.

wrt cathode. This of course occurs on positive alternations of the input signal.

The grid current is used to provide bias in the grid-leak-biasing system. As shown in Figure 7.9(a), the grid current charges the coupling capacitor. There can be no grid current in the opposite direction because the grid can collect electrons (when positive) but cannot emit them. Coupling-capacitor discharge current must, therefore, be through the grid resistor. This discharge current produces a voltage drop across the grid resistor. After the first few cycles of the input signal, a stabilized grid wrt cathode voltage waveform appears. It is shown as Figure 7.9(c). This grid wrt cathode voltage waveform [Figure 7.9(c)] has a negative average (dc) component. The dc portion of the waveform is the bias voltage for the tube. Note that the tube would have no bias if no input signal were present. Note also that the grid conducts on the positive peaks, which almost certainly produces distortion at the output during this interval of grid conduction. This circuit has seen considerable usage in old radio equipment.

CONTACT BIAS

As electrons from the space-charge region near the cathode move toward the plate of a vacuum tube, a small number of them may strike the grid structure and stick to it. For a circuit connected as in Figure 7.10, the only path for these electrons to take is through the

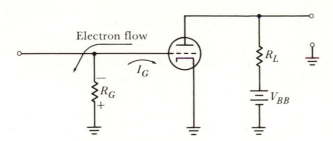

Figure 7.10   Contact bias.

grid resistor. This electron movement constitutes a current. The current is shown in terms of electron flow and conventional current. The magnitude of the current is very small. Therefore, to get a significant voltage drop across the grid resistor, the resistance must be very large. If, for example, the grid current is 0.1 μA and the grid resistor is 10 MΩ, the bias is 1 V.

SELF-BIAS (CATHODE BIAS)

The self-bias technique is quite commonly used in vacuum-triode circuits. Since the procedure is exactly the same as for FETs, you are referred to the FET chapter for details.

7.7 BASIC-AMPLIFIER CONNECTION: EQUATIONS AND WAVEFORMS

The equations and waveforms associated with the vacuum-triode version of the basic-amplifier connection are similar to those for FETs. Remember, however, that the vacuum tube is analogous to the N-channel depletion-mode FET. Refer to the FET chapter for further study.

7.8 AC EQUIVALENT CIRCUITS FOR VACUUM TRIODES

The ac equivalent circuits we choose to use for vacuum tubes are similar to those we used for FETs. Thus we may use either the voltage-source equivalent or the current-source equivalent. Refer to the FET chapter for details.

7.9 OTHER VACUUM-TUBE AMPLIFYING DEVICES

Several other types of vacuum-tube devices exist. They differ from the triode in the number of grids placed between the cathode and the plate. Each grid has a specific name. The grid nearest the cathode is called the *control grid*. The grid in a triode is a control grid.

The tetrode vacuum tube has two grids—the control grid and the screen grid. The screen grid is normally connected to a dc supply voltage somewhat less than the plate voltage. The screen grid is meant to reduce the amount of interelectrode capacitance between plate and control grid, that is, the input and output circuits. While the screen grid does reduce interelectrode capacitance, it also produces unusual *I-V* characteristics. The tetrode is not widely used because of its unusual characteristics.

A pentode is a vacuum tube with three grids. The control grid and screen grid are as in the tetrode. A pentode has a third grid, the suppressor grid, placed between the screen grid and the plate. The

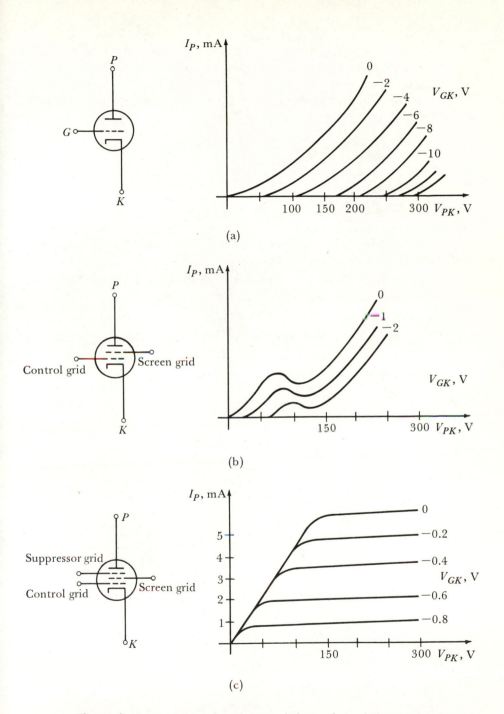

Figure 7.11 Vacuum-tube characteristics and symbols: (a) triode; (b) tetrode; (c) pentode.

suppressor grid effectively eliminates the undesirable characteristics that the tetrode displays. The pentode is a commonly used vacuum tube. It has high values of μ, g_m, and r_p.

The schematic symbols for, and the *I-V* characteristics of, each tube type are shown in Figure 7.11.

EXERCISES

QUESTIONS

Q7.1 Why is the invention of the vacuum triode considered a very important discovery?

Q7.2 What elements in a vacuum triode are analogous to the drain, the source, and the gate in an FET?

Q7.3 What is the function of a filament in a vacuum triode?

Q7.4 What is meant by a directly heated cathode?

Q7.5 What is the function of the control grid in a vacuum triode?

Q7.6 What parameters are used to characterize a vacuum triode mathematically?

Q7.7 Why is a vacuum triode more nearly analogous to an FET than a BJT?

CHAPTER 8
SELECTED CIRCUITS AND COMPONENTS

In this chapter we take up a diverse list of topics. Sections 8.1, 8.2, and 8.3 deal with transistors and tubes in other than the basic-amplifier configuration. These circuit connections are dealt with here because, although quite useful under some circumstances, they are of notably less importance than the basic-amplifier circuit upon which we have concentrated up to this point.

Sections 8.4 and 8.5 are concerned with SCRs and UJTs. The functions performed by, and the usefulness of, these devices is distinctly different from the analog-signal-amplification function of transistors and/or vacuum tubes. They can be considered as controlling or switching devices.

Integrated circuits are the main thrust of Sections 8.6 through 8.12. Integrated circuits are widely used in a variety of applications. An introduction to various integrated circuits is quite appropriate at this point, since we now have a good grasp of the capabilities and function of individual transistors.

Oscillator circuits are considered in Section 8.13. This coverage is nearly devoid of mathematical analysis. Hopefully it will be quite helpful in establishing certain concepts in the reader's mind. A mathematical analysis has not been done because there seems to be no appropriate stopping point without a complete coverage, which is beyond our needs at the present time.

271

8.1 COMMON-COLLECTOR, COMMON-DRAIN, AND COMMON-PLATE CONNECTIONS

The main thrust of this book has been on amplifier circuits which we have chosen to call basic-amplifier circuits. The basic-amplifier circuit is in fact the most widely used circuit whether we are talking about BJTs, FETs, or vacuum tubes. The basic-amplifier circuit is easy to understand. We have purposefully withheld other possible connections so that concentration could be on one type of circuit connection that is utilized in, perhaps, 90% or more of all single-stage amplifier circuits. The circuit we have called the basic-amplifier connection is technically called the common emitter, the common source, or the common cathode, depending on whether the active device is a BJT, an FET, or a vacuum tube. Before you proceed, make sure you thoroughly understand the basic-amplifier circuit.

In Figure 8.1 we have schematic drawings of a common-collector amplifier circuit, which is frequently called an emitter follower. In Figure 8.1(a) the circuit diagram is shown in a preferred form. In part (b) the schematic is arranged so that it is easy to see that the collector is the common element. In part (c) we have the ac equivalent of an emitter-follower circuit. It should be noted that the collector has a common connection with the input and the output.

The emitter follower we have chosen uses the universal-bias technique. Any of the bias techniques used for the BJT basic-amplifier circuit will also work for the emitter follower.

The common-collector circuit has features we should carefully note. The voltage gain will always be near unity but always less than unity. A value of 0.95 is typical. There is no polarity inversion between input and output signals. An emitter-follower circuit can be built using either an *NPN* or a *PNP* transistor. You must use the supply-voltage polarity appropriate for the transistor type you choose. The previous comments about voltage gain and relative polarity apply equally to *NPN* or *PNP* circuits. Other features of this connection will be considered near the end of this section.

An FET amplifier, in the common-drain configuration, is shown in the schematics of Figure 8.2. In part (a) the schematic is shown in a preferred arrangement. The form shown in part (b) tends to emphasize the common-drain connection. The ac equivalent of the circuit of parts (a) and (b) is shown in part (c), wherein it is especially easy to see that the drain is common to the input and output circuits. The common-drain configuration is frequently called a source follower.

The source-follower circuit is analogous to the emitter-follower

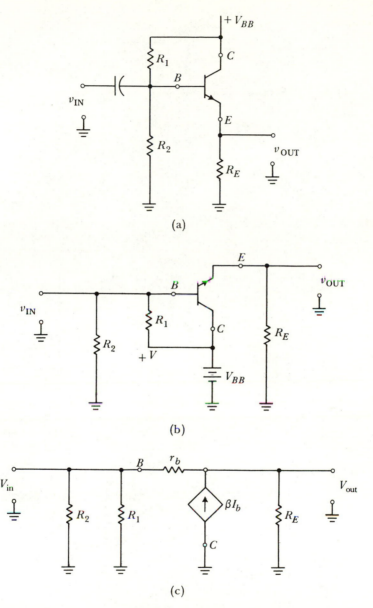

Figure 8.1 Common-collector (emitter-follower) amplifier circuit: (a) preferred configuration; (b) form emphasizing common-collector connection; (c) ac equivalent circuit.

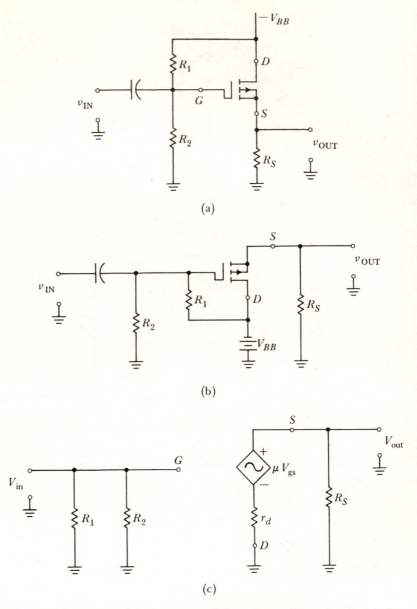

Figure 8.2 Common-drain (source-follower) amplifier circuit. (a) Preferred diagram configuration; (b) form emphasizing common drain; (c) ac equivalent circuit.

circuit previously discussed. As one would expect, a source follower will always have a voltage gain of less than unity. There is no polarity inversion associated with a common-drain amplifier circuit.

An FET is often considered a solid-state vacuum tube. Admittedly the two devices are analogous in many ways. In order to show the versatility of an FET, we have chosen a P-channel enhancement-mode FET for our source-follower example. By analogy, a vacuum tube is available only as an N-channel depletion-mode device. Even though we have chosen examples to point out the differences between vacuum tubes and FETs, note the similarities between the ac equivalent circuits of Figure 8.3(b) and Figure 8.2(c).

Figure 8.3 shows three schematic drawings of the same common-plate amplifier circuit. The configuration in part (a) emphasizes why the circuit is called a common-plate circuit. Part (b) is the ac equivalent of part (a). In part (b) the reason for the name is even more apparent. Note in part (b) that the input is applied between grid and plate terminals. The output is taken between cathode and plate. The plate terminal is thus common to both the input and output signals. In part (b), the ac equivalent circuit, the plate is connected to the circuit common or ground; thus this circuit is occasionally called a grounded-plate amplifier. In part (c) we have the preferred schematic configuration for a common-plate amplifier circuit.

The circuit of Figure 8.3 utilizes self-bias. Thus with no signal input, some dc plate (and cathode) current will flow. This dc current will provide a dc voltage across R_K. This dc voltage across R_K is the bias voltage for the tube. The output voltage v_{OUT} will have a dc level due to the bias voltage. The signal voltage (ac) will be superimposed upon the bias value. If the dc level at the output is to be eliminated, a coupling capacitor may be used.

The voltage gain of a cathode follower is always slightly less than unity. There is no polarity inversion between input and output. Since the cathode (output) signal has the same polarity and nearly the same ac signal level as the input, this circuit is commonly called a *cathode follower*.

We have referred to each circuit in this section as an amplifier circuit. You may wonder about the appropriateness of the word amplifier in view of the fact that the signal-voltage output is less than the signal-voltage input for each of the circuits. We think the word amplifier is appropriate even though the numerical value of the voltage gain might be $A_V = 0.9$, for example. A more thorough study of the common-collector, the common-drain, and the common-plate circuits would show that the current gain and/or the power gain of these circuits may be relatively large. A current gain of

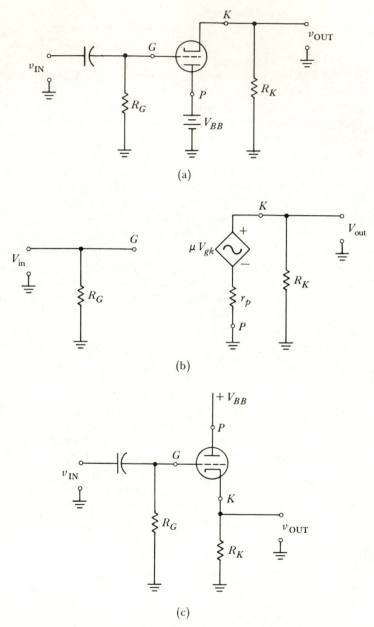

Figure 8.3   Common-plate (cathode-follower) amplifier circuit. (a) Form emphasizing common-plate connection; (b) ac equivalent of (a); (c) preferred schematic configuration.

several hundred is quite possible for a BJT in an emitter-follower circuit.

Students sometimes question why they should study a circuit whose voltage gain is less than unity. The question is quite reasonable. The following example might be helpful. Consider the amplifier stage that is to drive a loudspeaker in a radio. A speaker is a device that requires relatively large amounts of current to operate properly. The voltage requirements are moderate. In view of these facts, the stage that drives a speaker generally has a high current gain or power gain but may in all probability have a low voltage gain.

Perhaps it would be beneficial to pause and explain why the follower circuits of this section cannot have a voltage gain of greater than unity. The explanation is quite easy in the case of the emitter follower.

You will recall that in BJTs the emitter-base junction is always forward-biased. We sometimes make the assumption that the turn-on voltage for a silicon PN junction is a constant of 0.6 V. If this were the case, we would find that the emitter and base voltages would be as shown in Figure 8.4. If there is a constant 0.6 V

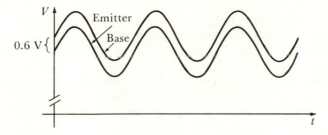

Figure 8.4   Waveforms for NPN emitter-follower circuit.

difference in dc level between the two waveforms, then the waveforms must be of the same peak-to-peak value. Since the base voltage represents the input-signal voltage waveform and the emitter voltage represents the output-signal voltage waveform, we see that we have equal input- and output-signal voltage levels. Thus the voltage gain is (near) unity.

In practice we find that the emitter-base voltage is not quite constant. This means that the emitter and base waveforms are not exactly the same amplitude and the gain of the stage is slightly less than unity. If you are not yet convinced, consider the results if somehow the base (that is, input) voltage should increase and the emitter (that is, output) voltage should increase a somewhat greater amount. Such circumstances would cause the base wrt emitter voltage to be somewhat less than 0.6 V. With the lesser base wrt emitter voltage, the base-emitter junction would tend to turn off and we would have less base current, less collector current, less emitter current, and thus less emitter voltage, that is, output voltage. We hypothesized that the emitter voltage somehow increased a greater amount than did the base voltage. We found that the result would be that the emitter voltage (output voltage) would decrease back toward a lower value. Hopefully you are convinced that the voltage gain of any of the follower circuits is less than unity.

The follower-type circuits can be used to advantage in several circumstances. The current and power gain can be very high, even approaching infinity for FETs and vacuum tubes. If you need current or power gain, consider these circuits. Follower circuits have a low output impedance. This feature may be very important in some applications. Although follower-type circuits are used less often than are basic-amplifier circuits, these circuits are important and should be understood thoroughly.

EXAMPLE 8.1   Emitter-follower circuits are quite commonly used. As mentioned before, they are often used because of their desirable impedance levels, which is why it is desirable to stop at this point to calculate the impedance levels for an emitter follower.

The circuit we will be considering is shown in Figure 8.5(a). The ac equivalent of the circuit is shown in part (b).

The input current is seen to be the sum of the currents through the bias resistors plus the base current. The input impedance of the circuit, that is, the load impedance seen by the signal source V_{in} is

$$Z_{in} = \frac{V_{in}}{I_{in}} = \frac{V_{in}}{I_{R1} + I_{R2} + I_b}$$

I_{R1} and I_{R2} can be determined by inspection. The base current I_b is found by writing a loop equation around the loop including V_{in}, r_b, and R_E.

$$V_{in} = I_b r_b + I_b R_E + \beta I_b R_E$$

$$= I_b(r_b + (\beta + 1)R_E)$$

Solving for I_b we obtain

$$I_b = \frac{V_{in}}{r_b + (\beta + 1)R_E} = \frac{V_{in}}{52.51 \text{ k}\Omega}$$

Note that the term $r_b + (\beta + 1)R_E$ is the input impedance of the

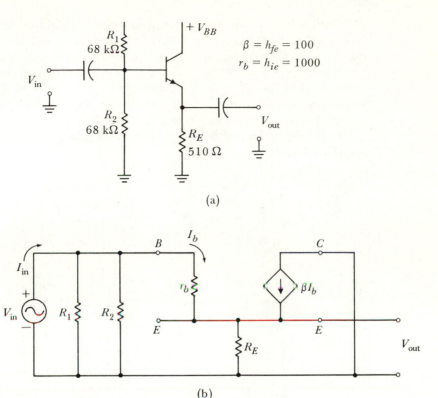

Figure 8.5 Emitter-follower (common emitter) circuit for Example 8.1. (a) Schematic diagram; (b) ac equivalent circuit.

emitter follower without considering the bias resistors. Collecting information for the Z_{in} equation, we obtain

$$Z_{in} = \frac{V_{in}}{\dfrac{V_{in}}{68 \text{ k}\Omega} + \dfrac{V_{in}}{68 \text{ k}\Omega} + \dfrac{V_{in}}{52.51 \text{ k}\Omega}} = 20.64 \text{ k}\Omega$$

Referring to Figure 8.5(b) we note that the input impedance is the parallel combination of R_1, R_2, and the input impedance of the basic emitter follower. The calculations show that in this case the input impedance is reduced from 52.51 kΩ to 20.64 kΩ by the bias resistors. When designing for high Z_{in}, then, we need to be careful in selecting appropriate bias levels and bias resistors. Also note that a higher input Z will result if we use a large value for R_E and pick a BJT with high β.

We will solve for the output impedance by methods similar to those used in Chapter 5. We will disable the normal signal-input source and apply a signal-input source at the (former) output terminals, and thereby find the "input" impedance at the (former) output terminals. The complete ac equivalent circuit is shown in Figure 8.6(a). Since the disabled V_{in} source shorts out the bias resistors, we have redrawn the circuit in a convenient way in part (b).

The circuit of Figure 8.6(b) is easily solved by writing a node equation, namely,

$$I_1 = I_{RE} - I_b - \beta I_b$$

$$= \frac{V_1}{R_E} - \left(\frac{-V_1}{r_b}\right) - \left(\frac{-\beta V_1}{r_b}\right)$$

$$= V_1\left(\frac{1}{R_E} + \frac{\beta + 1}{r_b}\right)$$

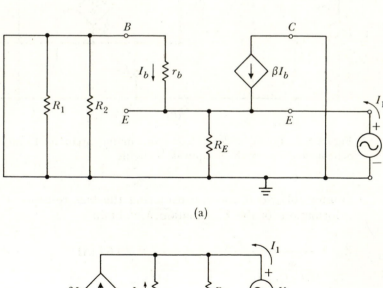

(a)

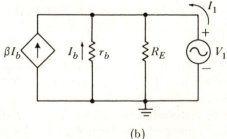

(b)

Figure 8.6 ac equivalent circuit for calculating output Z of emitter follower. (a) Original ac equivalent circuit; (b) simplified redrawn equivalent circuit.

The strange situation regarding the signs associated with the I_b and βI_b terms is due to the fact that V_1 is of such a polarity as to cause the current flow through r_b to be opposite the defined direction of I_b. Rearranging the equation, we find the output impedance to be

$$Z_{out} = \frac{V_1}{I_1} = \frac{1}{\dfrac{1}{R_E} + \dfrac{\beta + 1}{r_b}}$$

$$= \frac{R_E r_b}{r_b + R_E(1 + \beta)}$$

Substituting the numbers for our specific example, we have

$$Z_{out} = \frac{(500)(1000)}{1000 + (510)(101)} = 9.52 \ \Omega$$

The result we have just found is quite significant. It shows that the output impedance of an emitter follower is usually much lower than the emitter resistor. The low output Z feature is the main reason for the use of follower-type circuits. Since they have low output impedance, they are capable of driving loads of relatively low impedance.

8.2 COMMON-BASE, COMMON-GATE, AND COMMON-GRID CONNECTIONS

The amplifier-circuit connections to be discussed in this section are analogous to one another. In each case the common element is the control element. At least it is common and convenient to think of the base, the gate, and the grid as the respective control elements.

A common-base (CB) amplifier circuit is represented in Figure 8.7. In parts (a) and (b) we have the same circuit connection but the schematic is configured quite differently. The configuration of part (b) is intended to make it quite apparent that the base of the transistor is at ac ground (due to the near-zero X_c of the capacitor C) and that it (the base connection) is therefore common to the input and output terminals.

In Figure 8.7(c) we have the ac equivalent circuit of the common-base circuit. Note that the base is at ac ground. It is apparent that the input-signal voltage V_{in} is applied between emitter and base and the output-signal voltage is taken between collector and base. Thus the base is definitely common to both input and output circuits.

It is interesting to note that the capacitor that causes the base to be

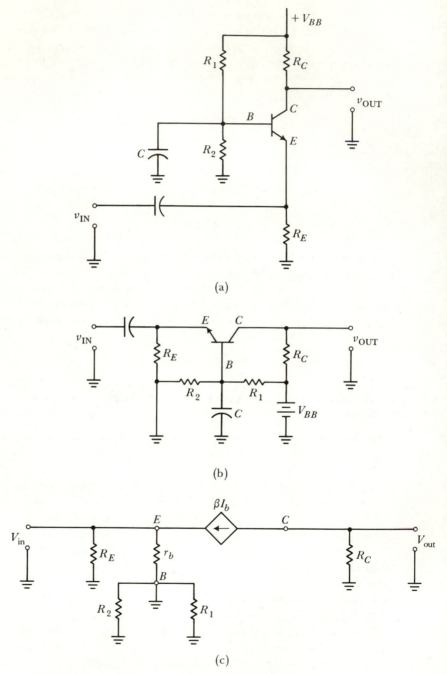

Figure 8.7 Common-base (CB) amplifier circuits (*N*-channel BJT): (a) and (b) alternate schematic configurations; (c) ac equivalent circuit.

at ac ground also causes R_1 and R_2 to have no effect on the ac operation of the common-base circuit. Even though these resistors appear in the ac equivalent circuit of Figure 8.7, they do not influence the operation of the circuit.

The FET version of the common-control-element amplifier circuit is shown in Figure 8.8. Specifically it is a common-gate amplifier circuit. We have chosen an N-channel insulated-gate FET for this example. The FET is assumed to be of the enhancement-mode variety and is biased via the voltage-divider method.

Figure 8.8(a) and (b) shows two schematic configurations of the same circuit. The resistors R_1 and R_2 are the voltage-divider resistors that provide the necessary bias voltage. The capacitor C is necessary to ensure that the gate is at ground potential with respect to the ac signal. In part (c) we have the ac equivalent of the common-gate amplifier circuit. Note the presence or absence of R_1, R_2, and C in the ac equivalent circuit. The resistors R_1 and R_2 have no effect on the circuit operation.

A common-grid amplifier-circuit connection is shown in Figure 8.9. This circuit is sometimes referred to as a grounded-grid connection. In parts (a) and (b) we have different schematic configurations of the same circuit connection. In part (c) we have the ac equivalent of the circuit. The input-signal voltage is applied between cathode and ground. Since the grid is at ground potential, we see that the input signal is effectively applied between the grid and cathode terminals.

The grounded-grid amplifier in Figure 8.9 employs self-bias. Fixed bias could be used if desired.

Let us consider some of the characteristics of the amplifier circuits presented in this section. Each of the circuits is capable of a high voltage gain. The current gain of each circuit is unity or slightly less. The power gain may be relatively high. The input signal and the output signal are of the same relative polarity. These characteristics would seem to describe circuits of great usefulness.

Amplifier circuits of the common-base, common-gate, and common-grid configurations are seldom used. Only in specialized applications are they utilized. In fact, these circuits have some characteristics that might be considered undesirable.

The main disadvantage of common-control-element circuits is their associated impedance levels. The impedance levels are just the opposite from what is normally desired. Whereas a high input impedance and a low output impedance are desirable, these circuits have a low input impedance and a high output impedance. The impedance characteristics make it very difficult to use a common-

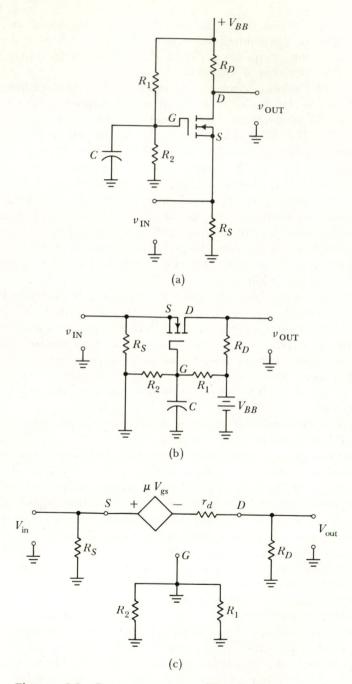

Figure 8.8 Common-gate amplifier-circuit connections (*N*-channel enhancement-mode IGFET): (a) and (b) alternate schematic configurations; (c) ac equivalent circuit.

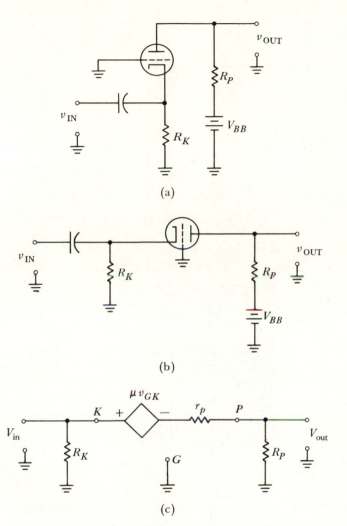

Figure 8.9 Common-grid amplifier-circuit connections: (a) and (b) alternate schematic configurations; (c) ac equivalent circuit.

control-element amplifier circuit successfully in a cascade connection.

An example should help the student understand why the impedance levels are not appropriate for a cascade connection. In fact, there are few situations where the impedance levels are appropriate. Figure 8.10(a) is a representation of a common-control-element amplifier circuit. The input impedance is shown as 10 Ω and the output impedance is shown as 1000 Ω. These values could be typical

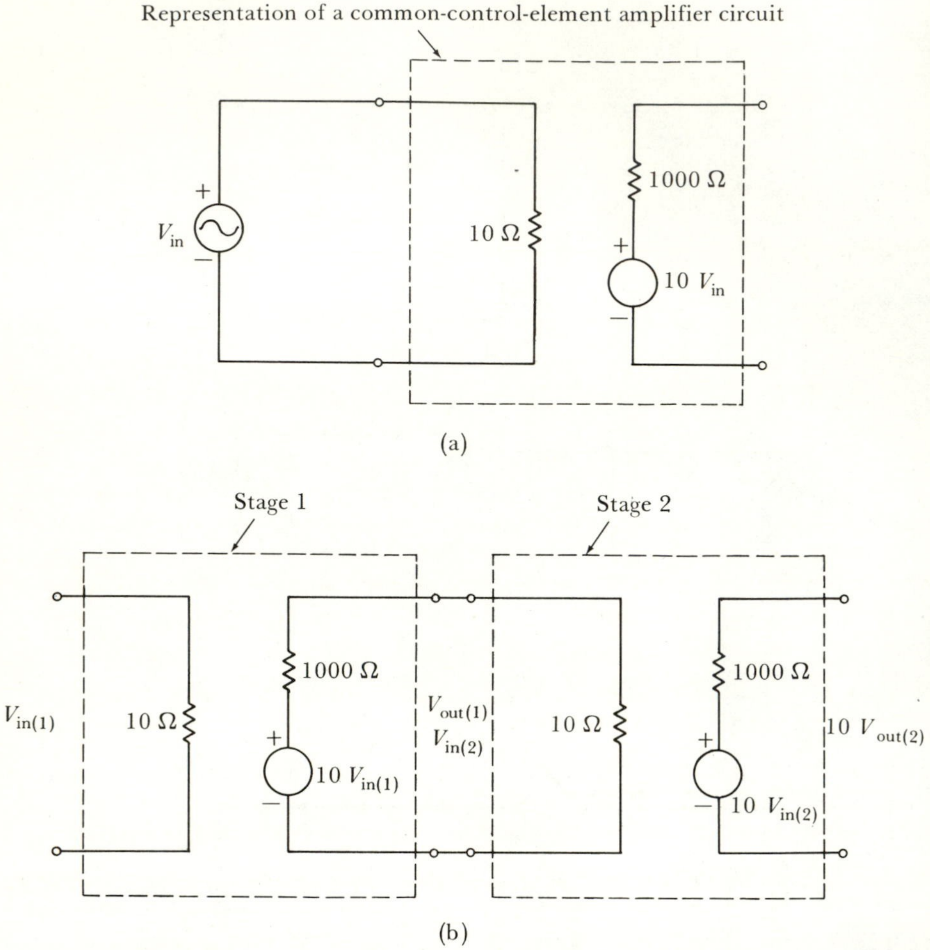

Representation of a common-control-element amplifier circuit

(a)

(b)

Figure 8.10 Common-control-element circuit representation: (a) single stage; (b) two-stage cascade connection.

for a common-base (CB) BJT amplifier circuit. The input is driven by a zero impedance source and the output terminals are unloaded. Under these conditions the input voltage is V_{in} and the output voltage is $10\ V_{in}$. The voltage gain is thus $A_V = 10$.

Figure 8.10(b) is a cascade connection of common-control-element amplifier circuits. In order to calculate the overall voltage gain of the cascade connection, we need to calculate the interstage voltage $[V_{out(1)} = V_{in(2)}]$ and the output voltage.

$$V_{\text{out (1)}} = V_{\text{in (2)}} = 10 \frac{V_{\text{in(1)}}(10)}{1000 + 10} \approx 0.099 V_{\text{in(1)}}$$

The interstage voltage is seen to be severely limited by the voltage-divider action of the input and output impedances. Since the second stage is unloaded, the output voltage is

$$V_{\text{out (2)}} = 10 V_{\text{in (2)}} = 0.99 V_{\text{in (1)}}$$

The overall voltage gain of the two-stage cascade connection is

$$A_V(\text{cascade}) = \frac{V_{\text{out(2)}}}{V_{\text{in(1)}}} = 0.99$$

Our arithmetic in this example has shown that the voltage gain of the two-stage cascade connection is slightly less than unity. This is so even though each stage has an unloaded voltage gain of 10. Normally we expect the gain of a cascade amplifier to be the product of the individual stage gains. In this two-stage example, we might have expected the overall voltage gain to be 100. It is actually less than unity. Obviously the impedance characteristics are important and in many cases are undesirable for common-control-element circuits.

Common-control-element circuits are occasionally used to advantage in high-frequency and low-noise applications.

8.3 DARLINGTON, OR SUPER BETA, TRANSISTOR CONNECTION

The Darlington connection of BJTs provides us with a convenient way of making one transistor out of two. It is a widely used connection and in certain applications offers significant advantages.

Figure 8.11 shows the Darlington connection of two transistors. In part (a) we can see the basic concept involved: the base current of one transistor (Q_2) is the emitter current of another transistor (Q_1). The significant advantage of a Darlington connection is that the resultant Darlington transistor has very high β. A little bit of thought will reveal that the overall β of a Darlington connection is

$$\beta_{\text{Darlington}} = \beta_{Q1}\beta_{Q2}$$

If both Q_1 and Q_2 have a β of 100, and if the collector current of Q_2 is, for example, 10 mA, then the following circumstances are valid.

$$I_{BQ2} = \frac{I_{CQ2}}{\beta_{Q2}} = \frac{10 \text{ mA}}{100} = 0.1 \text{ mA}$$

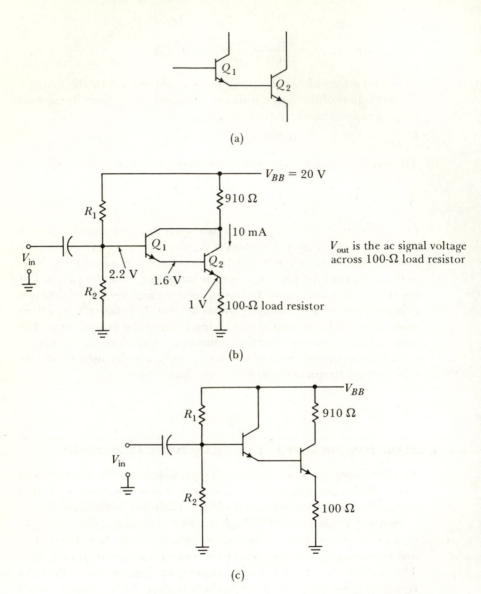

V_{out} is the ac signal voltage across 100-Ω load resistor

Figure 8.11 Darlington transistor connections: (a) basic Darlington connection; (b) basic amplifier circuit with Darlington transistor; (c) alternate way of connecting circuit of part (b).

$$I_{BQ2} = I_{EQ1} \approx I_{CQ1} = 0.1 \text{ mA}$$

$$I_{BQ1} = \frac{I_{CQ1}}{\beta_{Q1}} = \frac{0.1 \text{ mA}}{100} = 0.001 \text{ mA}$$

The β for a Darlington connection relates the collector current of Q_2 to the base current of Q_1. Thus

$$\beta_{\text{Darlington}} = \frac{I_{CQ2}}{I_{BQ1}} = \frac{10 \text{ mA}}{1 \text{ }\mu\text{A}} = 10,000$$

$$= \beta_{Q1}\beta_{Q2} = (100)(100) = 10,000$$

In Figure 8.11(b) we have a basic-amplifier circuit containing a Darlington transistor. We will calculate values of R_1 and R_2 so that the amplifier will be biased at $I_C = 10$ mA. By doing this calculation and the following input-impedance calculation, we should see the real advantage of using Darlington-connected transistors. The voltage values shown are wrt ground and can be verified by the student. The transistors are silicon and have a base-emitter turn-on voltage of 0.6 V. We will assume that each transistor has a β of 50. Thus $I_{BQ2} = 0.2$ mA and $I_{BQ1} = 4$ μA. If we let $I_{R2} = 5I_{BQ1}$ for very good bias stability, we have

$$I_{R2} = (5)(4 \text{ }\mu\text{A}) = 20 \text{ }\mu\text{A}$$

$$R_2 = \frac{V_{R2}}{I_{R2}} = \frac{2.2 \text{ V}}{20 \text{ }\mu\text{A}} = 110 \text{ k}\Omega$$

$$R_1 = \frac{V_{R1}}{I_{R1}} = \frac{(20 - 2.2) \text{ V}}{(20 + 4) \text{ mA}} = 741.7 \text{ k}\Omega$$

$$\approx 750 \text{ k}\Omega$$

The input impedance of the circuit of Figure 8.11(b) is the impedance (load) that an ac input-signal source would feel. In this case the source would see (ac equivalentwise) R_2 in parallel with R_1 in parallel with the input impedance of Q_1. The input impedance of Q_1 is

$$Z_{\text{in Q1}} = \frac{V_{BQ1}}{I_{BQ1}} = \frac{2.2 \text{ V}}{4 \text{ }\mu\text{A}} = 550 \text{ k}\Omega$$

Thus the input impedance of the amplifier as seen by the signal-input source is

$$Z_{\text{in}} = R_2 \| R_1 \| Z_{\text{in Q1}}$$

$$= 110 \text{ k}\Omega \| 750 \text{ k}\Omega \| 550 \text{ k}\Omega$$

$$\approx 81.7 \text{ k}\Omega$$

The input impedance calculated here is considerably higher than it would be without a Darlington-connected transistor.

In cases where a Darlington transistor has a capacitor bypassing the emitter resistance, the input impedance of the circuit is best considered using an ac-equivalent-circuit analysis. Since this is the case, we will take some time to develop an approximate ac equivalent circuit of a Darlington transistor.

In Figure 8.12(a) we have the ac equivalent circuits of two BJTs interconnected as in a Darlington connection. The base, emitter, and collector of each individual transistor are labeled with subscripts 1 and 2; and the base, emitter, and collector of the composite Darlington are labeled with a subscript letter D. In Figure 8.12(b) we have the approximate ac equivalent of the composite Darlington transistor.

Our purpose now is to show how the parameter values shown in Figure 8.12(b) are equivalent to, and based upon, the values shown

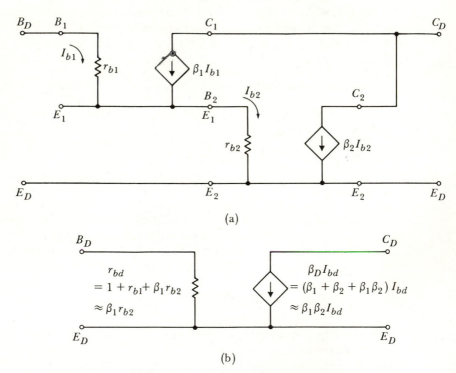

(a)

(b)

Figure 8.12 ac equivalent circuit of a Darlington transistor. (a) The interconnection of two individual ac equivalent circuits. (b) The composite ac equivalent circuit (approximate) of a Darlington transistor.

in Figure 8.12(a). In order to do this, we should make note of two simple facts that are apparent by inspection of part (a). First, the current into the Darlington collector terminal is

$$I_{CD} = I_{C1} + I_{C2}$$
$$= \beta_1 I_{B1} + \beta_2 I_{B2}$$

Moreover the current I_{B2} is really the sum of two other currents, that is,

$$I_{B2} = I_{C1} + I_{B1}$$
$$= \beta_1 I_{B1} + I_{B1}$$
$$= (\beta_1 + 1)I_{B1}$$

If we now substitute for I_{B2} in the equation for I_{CD}, we obtain

$$I_{CD} = \beta_1 I_{B1} + \beta_2(\beta_1 + 1)I_{B1}$$
$$= I_{B1}(\beta_1 + \beta_2 + \beta_1\beta_2)$$

Since $I_{B1} = I_{BD}$, we can write the equation as

$$\frac{I_{CD}}{I_{BD}} = \beta_1 + \beta_2 + \beta_1\beta_2$$

This last equation, which is collector current divided by base current for a Darlington connection, shows us that the β for a Darlington connection is

$$\beta_{\text{Darlington}} = \beta_1 + \beta_2 + \beta_1\beta_2$$

This equation is usually approximated, by omitting the first two terms, as

$$\beta_{\text{Darlington}} = \beta_1\beta_2$$
$$\beta_D = \beta_1\beta_2$$

This may appear to be a drastic approximation, but it introduces an error of less than 10% if $\beta_1 = \beta_2 = 20$. For higher β_1 and β_2 values, the error of approximation is less. The equations just derived for β_D shows why the dependent-current source of Figure 8.12(b) has the value shown.

Now we need to consider the resistance r_{bD} for the Darlington ac equivalent circuit. We note that this resistance is connected between base and emitter of the Darlington equivalent of Figure 8.12(b). Thus if we can find the voltage V_{beD} and the current I_{beD}, we can calculate the numerical value of r_{bD}.

The voltage V_{beD} is the sum of the voltages across r_{b1} and r_{b2}. The voltage across r_{b1} is

$$V_{rb1} = I_{B1}r_{B1}$$

The voltage across r_{b2} is

$$V_{rb2} = I_{B2}r_{b2}$$
$$= I_{B1}(1 + \beta_1)r_{b2}$$

Thus the sum of the voltages is

$$V_{beD} = V_{rb1} + V_{rb2}$$
$$= I_{B1}\, r_{b1} + I_{B1}\,(1 + \beta_1)r_{b2}$$
$$= I_{B1}(r_{b1} + r_{b2} + \beta_1 r_{b2})$$

By inspection we note that $I_{beD} = I_{B1}$. Thus we calculate the equivalent base resistance of the Darlington connection as

$$r_{bD} = \frac{V_{beD}}{I_{beD}} = \frac{V_{beD}}{I_{B1}}$$
$$= \frac{I_{B1}(r_{b1} + r_{b2} + \beta_1 r_{b2})}{I_{B1}}$$
$$= r_{b1} + r_{b2} + \beta_1 r_{b2}$$

For convenience, the numerical value of r_{bD} is approximated as

$$r_{bD} \approx \beta_1 r_{b2}$$

This approximation introduces less than 5% error in most cases.

We now have a method of finding the component values in the ac equivalent circuit for a Darlington-connected transistor. Thus the ac-equivalent-circuit analysis of any circuit containing a Darlington transistor can proceed as in any other ac-equivalent-circuit analysis.

The Darlington connection is commonly used in emitter-follower applications.

A Darlington connection really does nothing except increase the input impedance. However, this can be very important, especially when cascading stages. See Section 8.2 and Figure 8.10 for the problems associated with *not* having a high input impedance. See also Section 9.3 about cascading stages.

8.4 SEMICONDUCTOR CONTROLLED RECTIFIERS

The semiconductor controlled rectifier (SCR) is the most widely used member of a family of devices known as thyristors. Thyristors are four-layer (*PNPN*) devices. SCRs are commonly called *silicon controlled rectifiers*.

In Figure 8.13 we have two common representations of an SCR. In part (a) the emphasis is on the *PNPN* construction. In part (b) we have an imaginary partitioning of the device. The partitioning leads to part (c), where an SCR is represented as two transistors. The two-transistor representation of an SCR is common. It allows us to apply our understanding of BJTs in order to explain the characteristics and operation of SCRs.

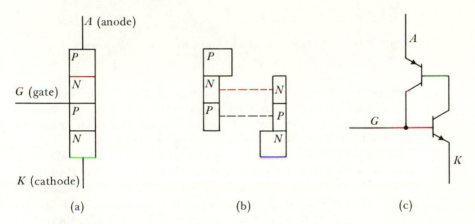

Figure 8.13   Semiconductor controlled rectifier (SCR): (a) *PNPN* (four-layer) representation; (b) transition; (c) two-transistor representation.

An SCR is a solid-state switch. It has two distinct modes of operation. In one mode it has extremely low resistance between anode and cathode terminals. In this mode, the *on* mode, the SCR closely resembles a switch whose contacts are closed or shorted. In the other mode, an SCR has very high resistance between the anode and cathode terminals. In this mode, the *off* mode, an SCR closely resembles a switch whose contacts are open.

The two modes of an SCR are illustrated in Figure 8.14. In part (a) we see the SCR, in the *off* mode, connected in a dc circuit. The voltages shown, both internal and external, are correct to a good

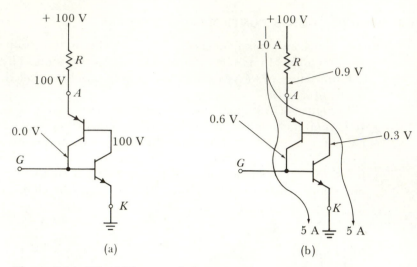

Figure 8.14   The SCR as a switch: (a) open switch; (b) closed switch.

approximation and are with respect to ground or the cathode terminal. Note that neither transistor has sufficient base wrt emitter voltage to cause it to turn on. This ensures that each transistor separately, and the SCR in total, is an open circuit.

In Figure 8.14(b) the SCR is acting as a closed switch. We have arbitrarily said that the current through the resistance R and the SCR is 10 A. The interior and exterior voltages as specified are approximately correct. Each transistor is conducting and is in fact "saturated." Note that the total current divides such that the collector of each transistor carries one-half the total current.

At this point we must pause to say something about saturation in transistors. Consider the circuit shown in Figure 8.15. The base current comes from some unspecified variable dc current source. As we have noted in earlier chapters, $V_{CE} = 10 - I_C R_L$. There is a lower limit to the value that V_{CE} can have. It is called the *saturation voltage*. V_{CE}(sat) is typically 0.3 V for silicon transistors and 0.1 V for germanium transistors.

Consider the circuit of Figure 8.15 as base current increases from zero. When $I_B = 0$, we find that $I_C = 0$ and $V_{CE} = 10$ V. As I_B takes on values larger than zero, I_C will take on values of βI_B, and V_{CE} will be decreasing in magnitude from 10 V. At a certain value of base current, the collector current (that is, βI_B) will be such that $V_{CE} = V_{CE}$(sat) and the transistor will be saturated. The amount of collector current required to saturate the transistor is dependent on the value

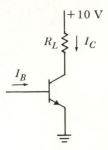

Figure 8.15 Transistor circuit to explain saturation.

of the collector resistor R_L. As base current increases past the saturating value, no significant changes occur in collector current or V_{CE} magnitudes. Base current can increase up to and beyond the collector-current value. Base current and collector current are no longer related by the normal β relationship. The only limitation on the value of base current is a maximum value specified by the transistor manufacturer. The limiting value is probably based on the power-dissipation capability of the transistor chip.

Now let us return to the SCR of Figure 8.14(b). Each transistor is well into the saturation region as indicated by the fact that base current is as large as collector current. Saturation explains why we know that the collector voltage of the *NPN* transistor is 0.3 V. The 0.6 V on the base of the *NPN* units is the normal *PN*-junction turn-on voltage. The values shown indicate that the SCR has 0.9 V across the anode-to-cathode terminals and is carrying 10 A of current. Under these conditions the *on* resistance of the device is 0.09 Ω. For very large current capacity SCRs, the *on* resistance might be as low as 1 mΩ.

The SCR of Figure 8.14(b) was assumed to be in the *on* state. Let us consider whether or not that assumption was valid. If the *NPN* transistor is in the *on* state, its collector current must be the base current of the *PNP* unit. With base current, the *PNP* unit will also have collector current. The collector current of the *PNP* unit is the same as the base current of the *NPN* unit. We have just proved that if the *NPN* unit has collector current, each transistor will be *on* and the entire SCR is in the *on* state.

Let us now consider the *I-V* terminal characteristics of an SCR. The *I-V* characteristic curves are typically of the shape shown in Figure 8.16. Note that as voltage increases from zero, the SCR allows very little current. The SCR is in the *off* state in this region. When a sufficiently high voltage is reached, we come to a break-over point.

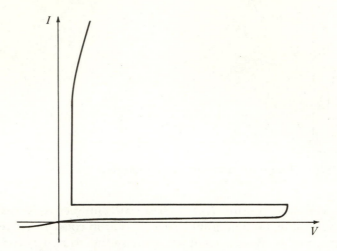

Figure 8.16   SCR *I-V* characteristics.

At that point we find that the SCR turns to *on* with the result that the current increases and the voltage across the SCR decreases drastically. This jump phenomenon is exactly what would occur in a circuit when a switch is suddenly closed.

We have now determined that the SCR is definitely a two-mode device. It makes an excellent switch. We now need to know how to open and close this solid-state switch. Once it is in the *off* state, it will stay in that state. Once placed in conduction, as in Figure 8.14(b), it will remain so indefinitely.

The *I-V* characteristics of Figure 8.16 give us a clue to one method of turning on an SCR. Any mechanism that we can think of that will increase the voltage across the device to the break-over point, even temporarily, will turn the device on. Although this is not normally the best way to turn on an SCR, there is a close relative of the SCR (the Shockley diode) that is intended to be operated in this manner.

The most common way to get an SCR into the *on* state is to make use of the gate terminal. If we apply a positive pulse of voltage at the gate wrt the cathode terminal such that there is a moderate amount of gate current, the *NPN* transistor will turn *on.* We have previously shown that if either transistor is turned to *on,* the entire SCR will turn *on.* Once the SCR is turned *on,* the gate voltage and current can be taken away and the SCR will remain in the *on* state.

Let us now consider how to turn *off* an SCR. Except for some recently announced members of the SCR family (the gate-turn-off SCR) we don't really have a method of making a conducting SCR suddenly turn *off*—at least not in the same sense that we can turn one *on*. If an SCR is placed in a circuit where the other circuit components cause the current to go to zero, the SCR will automatically revert to the *off* state. Actually the SCR will turn off at a current value known as the "holding current," which is a very small current compared with SCR rated current.

The turn-off procedure just discussed, that is, automatic self turn off as circuit current approaches zero, does not sound very satisfactory. As a matter of fact, the automatic self-turn-off feature of SCRs makes them quite adaptable to ac circuits. As a consequence, SCRs are commonly used in ac circuits and less commonly used in dc circuits.

Suppose that we would like to use an SCR to do a particular switching job that we envision. How will we choose a particular type from the hundreds of types produced? The answer is that we will have to look at the manufacturer's specifications. There are two ratings of particular importance: the blocking voltage, both forward and reverse, and the forward-current rating. If we plan to use the SCR with the standard 117-V ac line voltage, for example, we must pick an SCR with a breakdown voltage of at least a peak value of $\sqrt{2}$ × 117 V. If we fail to get a large enough breakdown voltage rating, the device may turn on in the forward direction without a gate pulse or it may break down in the reverse direction and thereby destroy itself. The forward-current rating must be at least equal in value to the average current through the device. Failure to get a large enough current rating will result in overheating and self-destruction.

There are many specifications of interest other than the primary quantities we have mentioned. The gate turn-on quantities are probably the most interesting of these remaining quantities, which should be self-explanatory from the data sheet. A typical manufacturer's data sheet is shown in Figure 8.17. Note the name of the device.

A simple application of an SCR acting as a switch is shown in Figure 8.18. Pay particular attention to the voltage waveshapes. Note that when the gate pulse is applied, the SCR turns on, that is, becomes a shorted switch, with the result that there can be no voltage across the SCR and there must be the full applied voltage across the load resistance. Also note that as the load current goes to zero (as indicated by v_{RL} going to zero), the SCR automatically returns to the *off* state, which means that the full applied voltage is across the SCR and none is across the load resistor until the next trigger pulse is applied.

- 25 A DC
- 50 V to 500 V
- 250 A Surge-Current
- Max I_{GT} of 20 mA

mechanical data

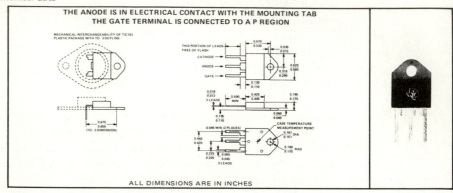

THE ANODE IS IN ELECTRICAL CONTACT WITH THE MOUNTING TAB
THE GATE TERMINAL IS CONNECTED TO A P REGION

ALL DIMENSIONS ARE IN INCHES

absolute maximum ratings over operating case temperature range (unless otherwise noted)

	TIC163F	TIC163A	TIC163B	TIC163C	TIC163D	TIC163E	UNIT
Repetitive Peak Off-State Voltage, V_{DRM} (See Note 1)	50	100	200	300	400	500	V
Repetitive Peak Reverse Voltage, V_{RRM}	50	100	200	300	400	500	V
Continuous On-State Current at (or below) 70°C Case Temperature (See Note 2)			25				A
Average On-State Current (180° Conduction Angle) at (or below) 70°C Case Temperature (See Note 3)			16				A
Surge On-State Current (See Note 4)			250				A
Peak Positive Gate Current (Pulse Width ≤ 300 μs)			3				A
Peak Gate Power Dissipation (Pulse Width ≤ 300 μs)			5				W
Average Gate Power Dissipation (See Note 5)			1				W
Operating Case Temperature Range			−40 to 110				°C
Storage Temperature Range			−40 to 125				°C
Lead Temperature 1/16 Inch from Case for 10 Seconds			230				°C

NOTES:
1. These values apply when the gate-cathode resistance $R_{GK} = 1 \text{ k}\Omega$.
2. This value applies for continuous d-c operation with resistive load. Above 70°C derate according to Figure 1.
3. This value may be applied continuously under single-phase 60-Hz half-sine-wave operation with resistive load. Above 70°C derate according to Figure 1.
4. This value applies for one 60-Hz half-sine-wave when the device is operating at (or below) rated values of peak reverse voltage and on-state current. Surge may be repeated after the device has returned to original thermal equilibrium.
5. This value applies for a maximum averaging time of 16.6 ms.

TEXAS INSTRUMENTS
INCORPORATED
POST OFFICE BOX 5012 • DALLAS, TEXAS 75222

Figure 8.17 Partial data sheet for a family of 1-A SCRs. (Courtesy of Texas Instruments, Inc.)

SERIES TIC163
P-N-P-N SILICON REVERSE-BLOCKING TRIODE THYRISTORS

electrical characteristics at 25°C case temperature (unless otherwise noted)

PARAMETER		TEST CONDITIONS		MIN	TYP	MAX	UNIT
I_{DRM}	Repetitive Peak Off-State Current	V_D = Rated V_{DRM},	R_{GK} = 1 kΩ, T_C = 110°C			2	mA
I_{RRM}	Repetitive Peak Reverse Current	V_R = Rated V_{RRM},	I_G = 0, T_C = 110°C			2	mA
I_{GT}	Gate Trigger Current	V_{AA} = 6 V,	R_L = 100 Ω, $t_{p(g)} \geqslant$ 20 μs			20	mA
V_{GT}	Gate Trigger Voltage	V_{AA} = 6 V, $t_{p(g)} \geqslant$ 20 μs,	R_L = 100 Ω, R_{GK} = 1 kΩ, T_C = −40°C			2.5	V
		V_{AA} = 6 V, $t_{p(g)} \geqslant$ 20 μs	R_L = 100 Ω, R_{GK} = 1 kΩ,	0.6	1.5		
		V_{AA} = 6 V, $t_{p(g)} \geqslant$ 20 μs,	R_L = 100 Ω, R_{GK} = 1 kΩ, T_C = 110°C	0.2			
I_H	Holding Current	V_{AA} = 6 V, T_C = −40°C	R_{GK} = 1 kΩ, Initiating I_T = 100 mA,			70	mA
		V_{AA} = 6 V,	R_{GK} = 1 kΩ, Initiating I_T = 100 mA			40	
V_{TM}	Peak On-State Voltage	I_{TM} = 25 A,	See Note 6			1.6	V
dv/dt	Critical Rate of Rise of Off-State Voltage	V_D = Rated V_D,	R_{GK} = 1 kΩ, T_C = 110°C		10		V/μs

thermal characteristics

PARAMETER		MAX	UNIT
$R_{\theta JC}$	Junction-to-Case Thermal Resistance	1	°C/W
$R_{\theta JA}$	Junction-to-Free-Air Thermal Resistance	36	

NOTE 6: This parameter must be measured using pulse techniques. t_w = 300 μs, duty cycle $\leqslant$ 2%. Voltage-sensing contacts, separate from the current-carrying contacts, are located within 0.125 inch from the device body.

THERMAL INFORMATION

AVERAGE ON-STATE CURRENT DERATING CURVE

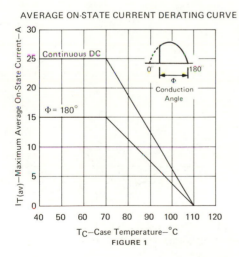

FIGURE 1

MAXIMUM CONTINUOUS ANODE POWER DISSIPATED
vs
CONTINUOUS ON-STATE CURRENT

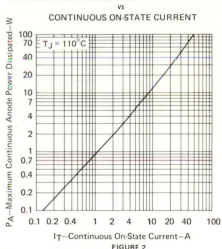

FIGURE 2

TEXAS INSTRUMENTS
INCORPORATED
POST OFFICE BOX 5012 • DALLAS. TEXAS 75222

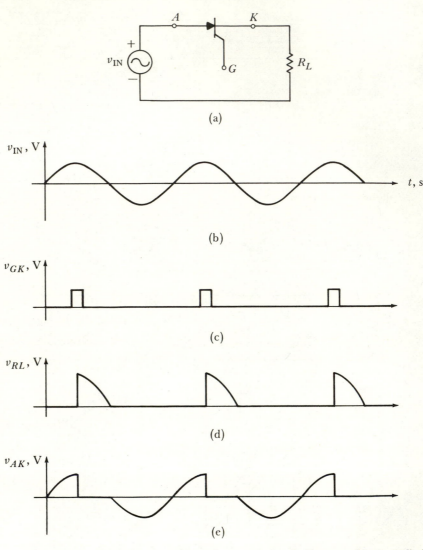

Figure 8.18 SCR circuit and waveforms: (a) circuit diagram; (b) ac-applied voltage waveform; (c) gate wrt cathode voltage waveform; (d) voltage across load resistance; (e) voltage across SCR.

The waveforms shown in Figure 8.19 should be helpful in explaining why we would want to use an SCR switch in an ac circuit. As we go from part (b) to (c) to (d) in Figure 8.19, the average power delivered to the load resistor increases. If R_L were a lamp, its brightness would increase from the (b) to (d) waveform. If R_L were a dc electric motor, such as an electric drill motor, its speed and torque would increase from (b) to (d). The waveforms given here show turn-

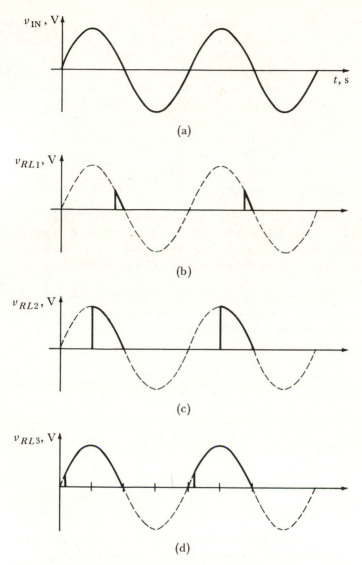

Figure 8.19   SCR circuit waveforms: (a) input-voltage waveform; (b) trigger late in alternation; (c) trigger at middle of alternation; (d) trigger early in alternation.

on during only one alternation. Turn-on during both alternations can be achieved by using two SCRs or one device equivalent to two SCRs (Triac).

Presumably you have noticed that we have thus far said nothing about the circuits that produce the gate pulse. Since the gating pulse is to be variable in position throughout the alternation, the trigger circuit required is really a phase-shifting circuit. There are several ways to generate the necessary signal for the gate of a SCR. Many of these circuits are rather tricky. Rather than attempt a comprehensive coverage of trigger circuits here, we refer the reader to the various SCR manuals or handbooks produced by RCA, Motorola, General Electric, Westinghouse, or other SCR manufacturers. One easily understood trigger circuit using a unijunction transistor is covered in Section 8.5.

8.5   UNIJUNCTION TRANSISTOR

The unijunction transistor (UJT) is quite different from either the BJT or the FET that we have studied in earlier chapters. It is so much different that the transistor part of its name may not be appropriate. Nevertheless, the device we will now look at is called a UJT.

A UJT is a three-terminal device. The terminals are called base 1, base 2, and emitter. The schematic symbol is shown in Figure 8.20(a), wherein the terminals are labeled appropriately.

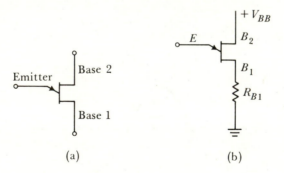

Figure 8.20   Unijunction transistor: (a) schematic symbol with terminals labeled; (b) partially connected UJT for explanation purposes.

In order to explain the characteristics of a UJT, consider the circuit of Figure 8.20(b). Between the B_1 and B_2 terminals we find a normal resistance of perhaps 10 kΩ. It is called the interbase resistance R_{BB}. Thus a small current flows from the voltage source, through the base terminals of the UJT, and through R_{B1}. Normally R_{B1} is very small in comparison with R_{BB}. Thus the current is almost entirely determined by R_{BB} rather than R_{B1}. The dc voltage across R_{B1} is negligible. If the voltage on the emitter terminal is gradually increased from zero, we find that the emitter current is near zero. (Actually it is equivalent in magnitude to the reverse current in a *PN*-junction diode.) As the emitter voltage is increased further, nothing happens until we reach a point called the emitter peak-point voltage or firing voltage. Once the emitter peak-point voltage is reached, the resistance between the emitter and base 1 suddenly drops to a very low value. A very large emitter current may occur under this condition. The emitter-to-base-1 resistance will remain at a very small value until either the emitter voltage or the emitter current reduces to a value called the holding, or extinguishing, value. When the holding value is reached, the UJT returns to its stable *off* state.

The emitter peak-point voltage V_P can be determined from the manufacturer's specifications. A reasonable approximation is

$$V_P = \eta V_{BB} + 0.5 \text{ V}$$

The parameter η (the Greek letter eta) is called the intrinsic standoff ratio. An examination of the specifications of many UJTs reveals that η is typically of a value between 0.5 and 0.8. V_{BB} is the supply voltage.

The UJT is commonly used in relaxation-oscillator circuits. Such a circuit is shown in Figure 8.21. The operation of the circuit is

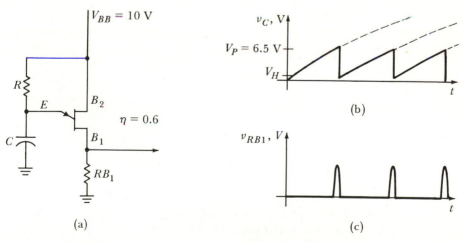

Figure 8.21 UJT relaxation oscillator: (a) circuit diagram; (b) voltage across capacitor; (c) voltage across R_{B1}.

explained quite easily. When the supply voltage V_{BB} is first applied, a current will flow in the *R-C* branch causing the capacitor to charge toward $V_{BB} = 10$ V. When v_C reaches the emitter peak-point voltage,

$$V_P = 0.6(10) + 0.5 = 6.5 \text{ V}$$

The UJT "fires" and very quickly discharges the capacitor through the low resistance between emitter and base 1. When the capacitor voltage reaches the holding value, the UJT extinguishes or returns to the *off* state. The cycle is repetitive. The waveforms of Figure 8.21(b) and (c) show that the holding voltage is very small and that the capacitor discharge current in R_{B1} produces a spiked-voltage waveform.

A UJT circuit is often used to trigger an SCR circuit. A simple example is the timer circuit shown in Figure 8.22. When switch S is closed, capacitance C will start to charge. When the emitter peak-point voltage is reached, the unijunction will fire and produce a voltage pulse across R_{B1}. The pulse across R_{B1} will trigger the SCR and thus energize the load. The load might be a lamp. This timer will return to the *off* mode when switch S is opened.

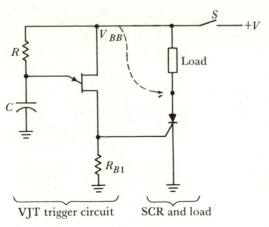

Figure 8.22   Timer circuit.

Consider the UJT trigger circuit of Figure 8.22 after the SCR has been turned on. Conditions around the UJT circuit have not changed, so it will continue to operate as a relaxation oscillator. Since the SCR is already in the *on* mode, the occasional trigger pulse will have no effect. If the designer wishes the unijunction circuit to be inoperative after the SCR fires, the alternate connection shown in dashed lines can be made. With this connection the UJT loses its supply voltage when the SCR fires.

For those of you who want to build the circuit of Figure 8.22, a few comments might be helpful. A likely mistake would be for the designer to try to choose R and C values to give a long time delay. If the R value gets too large, the leakage resistance of the capacitor may become significant and form a voltage divider with the resistor such that the capacitor voltage may never reach the emitter peak-point voltage. Another possible problem is that for very large values of R, the emitter leakage current may be enough to keep the voltage from reaching the peak-point value. Our comments thus far would seem to indicate that longer timer intervals should be obtained by choosing larger C values. Such is the case except that large-value capacitors may have higher leakage currents (that is, lower leakage resistance).

Another example of a UJT used as a trigger generator for an SCR is shown in Figure 8.23. The rectifier diode, the 10-kΩ 1-watt resistor, and the 22-V Zener diode form a dc power-supply circuit. The voltage across the Zener diode will be 22 V for most of the positive alternation of the ac input. Thus the UJT circuit has dc power only for the positive alternation. This situation is quite satisfactory, since the SCR is capable of turn-on only during the positive

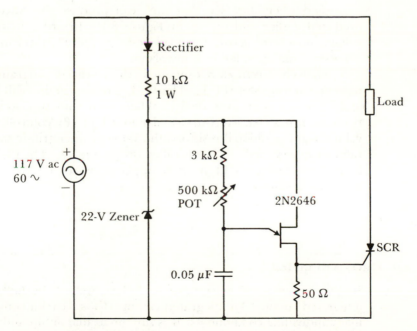

Figure 8.23 A practical circuit consisting of a UJT phase-shifting circuit to provide a trigger pulse for the SCR.

alternation. The UJT circuit is a standard relaxation oscillator. Each time it fires, a trigger pulse is applied to the gate of the SCR, thereby causing the SCR to turn on.

The 500-kΩ potentiometer in Figure 8.23 varies the R-C time constant of the charging circuit. When the resistance is small, the time constant is short, and the UJT fires early in the positive alternation. When the pot is adjusted for a large resistance, the time constant will be longer. Thus the UJT, and consequently the SCR, will fire late during the positive alternation.

The previous explanation has indicated that we can change the time instant, during the positive alternation, at which the SCR will be turned on. When the operator mechanically rotates the shaft of the potentiometer, the UJT responds by changing the phase angle at which the SCR will be turned on. The SCR conduction angle, and the power delivered to the load, may be conveniently controlled merely by adjusting the potentiometer setting. The circuit of Figure 8.23 is capable of shifting the firing point throughout most of the positive alternation.

If you want to experience an interesting educational exhibit, build the circuit of Figure 8.23. (Be careful of the rather high voltages involved.) Most students are pleasantly surprised and quite impressed when they see that they can produce the waveforms predicted earlier and shown in Figure 8.19. In order to see the voltage waveform across the load resistor, the 117VAC must be obtained from an isolation transformer.

A relatively recent development is the so-called programmable unijunction transistor (PUT). The PUT is related to the SCR family in that it is a four-layer device. When used with two additional resistors, it has characteristics similar to a UJT. By controlling the values of the additional resistors, the value of the intrinsic standoff ratio η can be varied. Thus it is called programmable. The PUT may be preferred over a UJT in some applications because of the programmable feature.

8.6 CMOS AMPLIFIER

In this section, and in sections to follow, we will be looking at integrated circuits. An integrated circuit (IC) is a circuit containing many individual components. It is unique in that all the individual components are formed on one semiconductor chip. Thus you may hear it called a *monolithic integrated circuit.*

We do not intend a detailed study of the internal construction of integrated circuits (ICs). We will be almost totally concerned with the terminal characteristics of ICs. Thus we will be considering an IC as a single component, an approach that is quite a simplification. It allows us to consider tens, hundreds, or even thousands of individual parts as a single multiterminal device.

The first integrated component we will consider is what we have chosen to call a CMOS amplifier. It is quite simple in that only two internal microscopic parts are interconnected to form the amplifier. This component has very interesting characteristics. The fact that it is very simple internally is an added attraction that should help to keep the purchase price reasonable.

The name CMOS is pronounced "see-moss." The letters stand for complementary metal-oxide semiconductor. The MOS part is exactly the same as that of the MOSFETs we have already studied. A CMOS amplifier is formed of two MOSFETs. The complement part of the name means that both *N*-channel and *P*-channel MOSFETs are used in the device. Their *I-V* characteristics are complementary in forming a very useful component.

The component we are calling a CMOS amplifier was originally manufactured and sold primarily as a digital-computer circuit. In the computer design field it would be called an inverter or perhaps an inverting buffer. This family of devices is known by several names. In addition to the name CMOS, individual manufacturers use their own identifying names. The RCA company uses the name COS/MOS, and Motorola Inc. uses the name McMOS. Several other companies also make this product.

In Figure 8.24 we have the proper connection of two MOSFETs to form a CMOS amplifier. Note that the top transistor is of the *P*-channel variety and the bottom one is an *N*-channel unit.

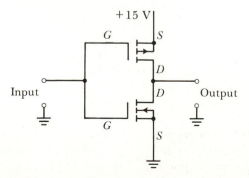

Figure 8.24 CMOS amplifier internal circuitry.

In order for the CMOS amplifier to work properly, each MOS-FET must be an enhancement-mode device. Each transistor is thus a normally-off unit. A review of the chapter pertaining to the fundamental operation of IGFETs may be advisable at this time. The review will remind us particularly of the proper voltage polarities for N-channel and P-channel devices and point out that insulated-gate devices have essentially zero current into the gate terminal.

Let us now consider the operation of the CMOS amplifier shown in Figure 8.24. Note that the output terminal is connected to a voltage-divider network between the positive-supply voltage and ground. The voltage-divider resistors are the channel resistances of the respective IGFETs. The output voltage can thus be anywhere between 0 and 15 V, depending on the respective values of the channel resistances of the voltage-divider network.

Let us consider the extremes first. If the input terminal is at the minimum value of zero, the upper (P-channel) transistor will be *on* and the lower (N-channel) transistor will be *off*. The upper voltage-divider resistor will therefore have low resistance and the lower voltage-divider resistor will have high resistance. The output terminal voltage will thus be near +15 V when the input terminal is at 0 V.

The other extreme condition is when the input terminal is at +15 V. Then we find that the upper transistor is off and the lower transistor is on. In this circumstance the lower voltage-divider resistor has a low resistance value and the upper voltage-divider resistor has a high resistance value. These conditions ensure that the output-terminal voltage will be near zero.

The extreme conditions on the input terminal have given us some knowledge of the operation of the device. We have found that the output-terminal voltage can vary between zero and the supply voltage. We note that the output terminal as compared with the input terminal is always at the opposite end of its voltage range. Since we have observed that extreme values on the input terminal cause the output terminal to take on limiting voltage values, it doesn't seem unrealistic to assume that some intermediate voltage on the input terminal will result in an intermediate value of output-terminal voltage. Such is in fact the case. If the P-channel and N-channel characteristics are properly matched, the output-terminal voltage will be at midrange when the input-terminal voltage is near midrange.

Several interesting features combine to make the CMOS amplifier very easy to bias. These features are extremely high input resistance, a polarity reversal between input and output terminals, and a good operating point obtained when both input and output terminals are at the same voltage. A single-resistor bias circuit is shown in Figure 8.25. This circuit is quite satisfactory for most applications. The

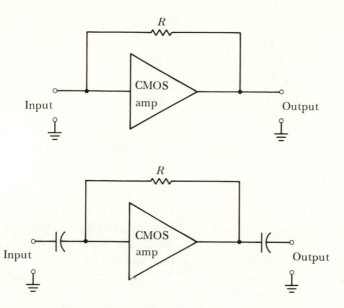

Figure 8.25 CMOS-amplifier-circuit bias technique: (a) biased amplifier; (b) amplifier with ac coupling to remove dc levels.

value of R is not critical. It may be anywhere from a few kilohms to many megohms. Since the input draws no current, there is no voltage drop across R. It merely serves to establish the same voltage on input and output terminals. Whereas a wider range of values will bias the amplifier, a value between 100 kΩ and 10 MΩ is probably desirable. The triangles in Figure 8.25 represent CMOS amplifiers. Only the signal-input and -output leads are shown.

The bias method of Figure 8.25 forces conditions to be as follows. There is no current through bias resistor R, and therefore the output and input terminals must be at the same voltage. This common voltage will be near one-half the supply voltage. Since the output terminal is at one-half the supply voltage, we know that both MOSFETs are conducting and have equal channel resistances.

As a method of proof, we will assume conditions different from the forced conditions just stated. We will discover contradictory facts that indicate that the assumed conditions are false.

Assume the following conditions:

1. Supply voltage = 15 V.

2. Common input-output voltage = 10 V (instead of one-half supply voltage).

As a consequence of the two assumed conditions, the following conditions result.

3. $V_{DS} = 5$ V for P-channel MOSFET (upper unit).

4. $V_{DS} = 10$ V for N-channel MOSFET (lower unit).

5. $V_{GS} = 5$ V for P-channel MOSFET.

6. $V_{GS} = 10$ V for N-channel MOSFET.

7. R_{DS} (N channel) $= 2R_{DS}$ (P channel).

For enhancement-mode devices, such as those in CMOS amplifiers, the channel resistance will decrease with increasing gate wrt source voltage. Conditions 5 and 6 indicate that the N-channel MOSFET has by far the greatest gate wrt source voltage. The N-channel unit should therefore have a lesser channel resistance than the complementary P-channel unit. Condition 7, however, indicates that the N-channel unit has twice as much channel resistance as does the P-channel unit. Conditions 5, 6, and 7 are definitely contradictory. The contradiction indicates one of two things: the assumed conditions are not true or the MOSFETs do not have complementary enhancement characteristics.

As the name indicates, the MOSFETs in a CMOS amplifier have complementary enhancement characteristics. The results of the preceding paragraph, then, must convince us that the assumed conditions were false. The only set of conditions that will not prove to be false are the conditions that the common input-output terminal voltage be one-half the supply voltage. Thus the bias circuit of Figure 8.25 does in fact force conditions as stated previously. The common input-output terminal voltage will be near one-half the supply voltage.

Now that we know how to bias a CMOS amplifier circuit, we need to consider its amplifying abilities. One way of doing this is to plot a graph of input voltage versus output voltage. Such a plot is shown in Figure 8.26. It is taken from the specification sheet for an RCA CA3600E. The operating point, that is, the Q point, is at one-half supply voltage, or 7.5 V. This plot thus represents a CMOS amplifier that uses the one-resistor bias technique we have discussed.

On the input-voltage axis of Figure 8.26, we have an ac signal voltage superimposed on the dc bias voltage. The ac signal applied to the input terminals causes a response at the output terminals. The response is plotted on the output-voltage axis of the graph. Since the response (ac output voltage) is larger than the excitation (ac input voltage), we conclude that the CMOS amplifier has a voltage gain larger than unity. For a CA3600E device, the small-signal voltage gain is about 15.

The input-output voltage plot of Figure 8.26 gives us information regarding linearity, and therefore the fidelity as an audio-signal

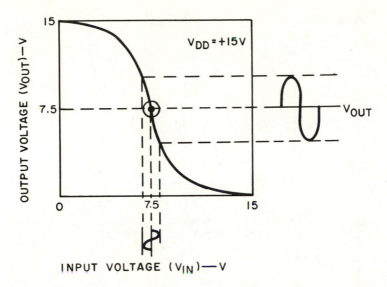

Figure 8.26 CMOS amplifier input-voltage–output-voltage transfer characteristics. (Courtesy of RCA.)

amplifier. The linearity, as indicated by a straight line on the input-output plot, is seen to be good near the bias point. As the ac signals get larger such that more and more of the transfer characteristic is utilized, the linearity decreases. For large signals the linearity gets quite poor. In the extreme case where the input signal is 15 V peak-to-peak, the output voltage will also be 15 V peak-to-peak and the fidelity will be very bad. We conclude, then, that the CA3600E is not suitable as an audio amplifier when signal levels approaching 15 V peak-to-peak are required.

At this point we should be well acquainted with CMOS amplifiers. We can think of many applications. In many of these cases we will want to use coupling capacitors to isolate the dc bias levels, as shown in Figure 8.25. This is particularly true when cascaded stages are involved.

The CMOS amplifiers we have been discussing are marketed by several companies. They are characterized for either analog or digital applications. Since this book discusses analog amplification only, we will first consider CMOS amplifiers intended for this application.

In Figure 8.27 we have CMOS amplifier circuits. In part (a) we have a schematic representation of the three CMOS amplifiers that are contained in one package. In part (b) we have the three stages connected in an ac-coupled cascade configuration. In part (c) we have a picture of the 14-lead dual in-line plastic package (DIP) in

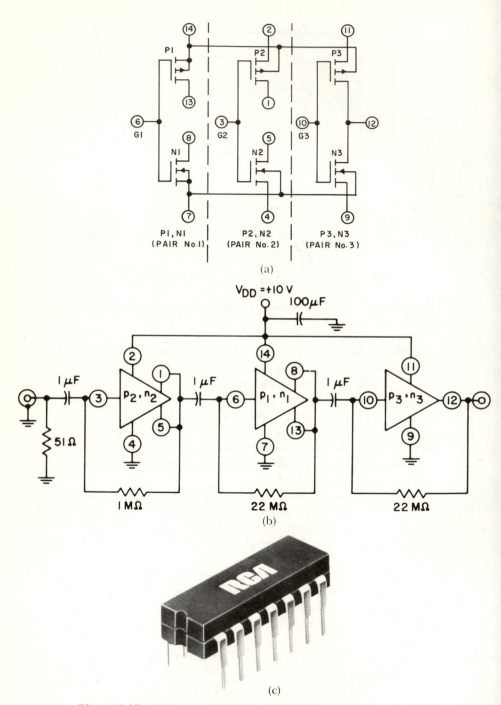

Figure 8.27 RCA type CA3600E COS/MOS transistor array: (a) schematic; (b) cascaded connection; (c) DIP package. (Courtesy of RCA.)

which the amplifiers are contained. Each part of Figure 8.27 is taken directly from an RCA specification sheet for their type CA3600E COS/MOS transistor array.

Several comments are in order regarding Figure 8.27. Some external package pins must be connected together in order to form CMOS amplifiers from the parts within a CA3600E package. For example, pins 1 and 5 are connected together in Figure 8.27(b). The bias resistors shown in part (b) are either 1 or 22 MΩ. This is not critical. Either value will work well in any location. The 1-μF coupling capacitors are not critical in value. The amplifier will have very good low-frequency response with these capacitor values. The 51-Ω resistor shown is intended as a termination for the input-signal generator. It has nothing to do with the operation of the cascaded CMOS amplifier. The 100-μF capacitor is a power-supply filter capacitor. It is not needed if there is a decent dc power supply. The power-supply voltage is shown as 10 V. Any voltage between 3 and 15 is satisfactory. The 15-V value is becoming a standard value.

CMOS amplifiers may also be constructed using circuits primarily characterized for digital applications. A prime example is the 74C line of CMOS ICs made by several manufacturers. In Figure 8.28 we have several applications using 74C-type devices taken directly from the *CMOS Integrated Circuits Data Book* published by National Semiconductor Corporation. In Figure 8.28(a) we have the package pinout assignments and in (b) we see the basic structure which constitutes each amplifier (inverter) in the 74C04 package. In (c) we see a biased amplifier. The resistor R_{in} is not needed unless it is desired to limit the gain of the amplifier. In part (d) we have a multistage amplifier. The 10-MΩ and 1-MΩ resistors will have meaning to you (as R_f and R_i, respectively) after you study operational amplifiers later in this chapter. In part (e) we see 74C-type devices used as a power-booster output amplifier for an operational amplifier. The oscillator circuit of part (f) should have meaning after a study of the oscillator section of this chapter. Even though several of these applications are premature at this point, they are included here to show the versatility of the simple little CMOS amplifier shown in Figure 8.28(b).

Some words of caution about using CMOS amplifiers. Any unused inputs should be connected to V_{BB} or to ground. Failure to do so will at the least waste power and at the worst destroy some types of devices. Some device types are buffered. They usually have a B suffix after the part number. These types may oscillate at a high frequency instead of operate as we have indicated in an amplifying mode. Some devices, for example the type CD4049, are intended for applications needing larger current-driving abilities. When used as

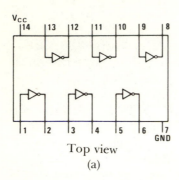

Top view
(a)

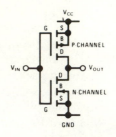

Basic CMOS inverter
(b)

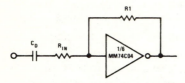

A 74CMOS invertor biased for linear mode operation

(c)

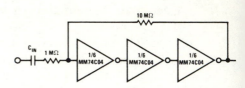

Three CMOS inverters used as an X10 ac amplifier

(d)

Phase-shift
oscillator using MM74C04
(f)

MM74C00 and MM74C02 Used as a Post
Amplifier to Provide Increased Current Drive.
(e)

Figure 8.28 CMOS applications: (a) package pinout; (b) basic internal structure; (c) single stage amplifier; (d), (e), (f) various applications. (Courtesy of National Semiconductor)

amplifiers, these types may dissipate excess power and/or self-destruct.

We have specifically mentioned National Semiconductor Corporation and RCA as producing CMOS devices. Each of these companies has a section on analog applications of CMOS devices in their literature. For further information please refer to one or both of the following publications: *CMOS Integrated Circuits,* a data book published by National Semiconductor Corporation; *RCA COS/MOS Integrated Circuits Manual,* published by Radio Corporation of America.

Hopefully we have learned several things in this section. First of all we have learned to consider as one component more than one device when properly interconnected on a single integrated-circuit semiconductor chip. We have refreshed our memory about FET devices. We have seen a unique bias circuit. Finally, we have seen a very interesting interconnection of complementary MOSFETs to form a very useful circuit.

8.7   INTEGRATED VOLTAGE REGULATORS

An integrated-circuit family of considerable interest and usefulness is the voltage-regulator family. Integrated voltage regulators are made in many configurations and by many manufacturers.

Before we look further at voltage regulators, we should discuss why voltage regulators are needed. It seems that the output voltage from most dc power supplies is not really constant. It probably changes somewhat as the amount of load current changes. Many times the output voltage is some function of temperature. The dc output voltage is also probably affected by any variation in the magnitude of the ac input voltage. Since the voltage output tends to change, we can profitably use a circuit or device to limit the amount of unwanted voltage change.

The voltage regulation of a circuit is a measure of how much the voltage changes under varying conditions. The load regulation, for example, is defined as

$$\text{load regulation} = \frac{V_{NL} - V_{FL}}{V_{FL}}$$

where V_{NL} = output voltage under no-load conditions

V_{FL} = output voltage under full-load (that is, rated-load conditions)

A power supply is loaded when it supplies current and power to an electric load. The full-load value is the maximum rated value of the power supply in question.

A close examination of the regulation equation will reveal that a small value of regulation is desirable in that it indicates a small change in output voltage as the load changes between no-load and full-load values. Regulation is often expressed as a percentage, in which case the value obtained from the regulation equation must be multiplied by 100%.

Integrated-circuit voltage regulators are components that are used to improve the voltage regulation of an existing power supply. The regulator is placed in the circuit between the dc power supply and the load. In Figure 8.29 we see how voltage regulators are used in practical situations. Note that the regulators regulate at a voltage less than the input voltage. Some minimum voltage drop across the regulator will be specified by the manufacturer. As indicated in Figure 8.29, single- or dual-polarity voltage regulators are available.

An integrated-circuit voltage regulator may significantly reduce the cost and/or size of a dc power supply. The reductions are due to the fact that the voltage-regulator circuit is quite efficient at removing ripple voltages. If it is known that a voltage regulator is going to be used, a less complex and less expensive main-power-supply filter circuit is satisfactory.

A special application for IC voltage regulators is as remote regulators. Consider the case where one large power supply is to provide a voltage source for several printed circuit boards. Even if the main

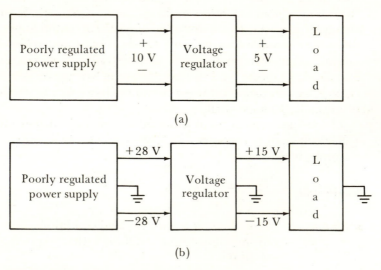

(a)

(b)

Figure 8.29   Voltage-regulator applications: (a)+5-V regulator; (b)±15-V regulator.

FIXED ±15V DUAL-TRACKING VOLTAGE REGULATORS

4195

DESIGN FEATURES

- ±15V Op-Amp power at reduced cost and component density
- Thermal shutdown at T_j = +175°C in addition to short-circuit protection

- Output currents to 100mA
- May be used as single output regulator with up to +50V output
- Available in TO-66, TO-99, and 8-pin plastic mini-DIP

The RM4195 and RC4195 are dual polarity tracking regulators designed to provide balanced positive and negative 15 volt output voltages at currents to 100mA. These devices are designed for local "on-card" regulation eliminating distribution problems associated with single point regulation. The regulator is intended for ease of application. Only two external components are required for operation (two 10μF bypass capacitors).

The device is available in three package types to accommodate various applications requiring economy, high power dissipation, and reduced component density.

SCHEMATIC DIAGRAM

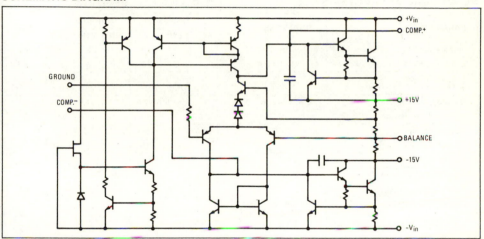

CONNECTION INFORMATION

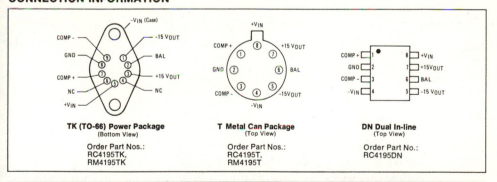

TK (TO-66) Power Package
(Bottom View)

Order Part Nos.:
RC4195TK,
RM4195TK

T Metal Can Package
(Top View)

Order Part Nos.:
RC4195T,
RM4195T

DN Dual In-line
(Top View)

Order Part No.:
RC4195DN

SEMICONDUCTOR DIVISION • 350 ELLIS STREET • MOUNTAIN VIEW, CALIF. 94040

Figure 8.30 Partial data sheet for an IC voltage-regulator circuit. (Courtesy of Raytheon Semiconductor.)

FIXED ±15V DUAL TRACKING VOLTAGE REGULATORS

ABSOLUTE MAXIMUM RATINGS

Input Voltage ±V to Ground .... ±30V
Power Dissipation @ T_A = +25°C
  TK Package 2.4W
  T Package 800mW
  DN Package 600mW
Load Current 150mA
Operating Junction Temperature Range .... RM4195: −55°C to +150°C. RC4195:   0°C to +125°C
Storage Junction Temperature Range .... RM4195: −65°C to +150°C. RC4195: −65°C to +125°C
Lead Temperature (Soldering, IO's) .... +300°C

ELECTRICAL CHARACTERISTICS (I_L = 1mA, V_{CC} = ±20V, C_L = 10µT unless otherwise specified)

PARAMETER	CONDITIONS	RM4195			RC4195			UNITS
		MIN	TYP	MAX	MIN	TYP	MAX	
Line Regulation	V_{in} = ±18 to ±30V		2	20		2	20	mV
Load Regulation	I_L = 1 to 100mA		5	30		5	30	mV
Output Voltage Temperature Stability			0.005	0.015		0.005	0.015	%/°C
Standby Current Drain	V_{IN} = ±30V,, I_L = 0mA		±1.5	±2.5		±1.5	±3.0	mA
Input Voltage Range		18		30	18		30	V
Output Voltage	T_j = +25°C	14.8	15	15.2	14.5	15	15.5	V
Output Voltage Tracking			±50	±150		±50	±300	mV
Ripple Rejection	f = 120Hz, T_j = +25°C		75			75		dB
Input-Output Voltage Differential	I_L = 50mA	3			3			V
Short-circuit Current	T_j = +25°C		220			220		mA
Output Noise Voltage	T_j = +25°C, f = 100Hz to 10KHz		60			60		µV RMS
Internal Thermal Shutdown			175			175		°C

THERMAL CHARACTERISTICS

PARAMETER	CONDITIONS	PACKAGE			UNITS
		DN	T (TO-99)	TK (TO-66)	
Power Dissipation	T_A = 25°C	0.6	0.8	2.4	W
	T_C = 25°C		2.1	9	
Thermal Resistance	ϕj-C		70	17	°C/W
	ϕj-A	210	185	62	

Figure 8.30 (continued)

voltage supply is perfectly regulated, the regulation at the printed circuit boards may be relatively poor due to voltage drops in the connecting wires. By placing an IC voltage regulator on each printed circuit card, the voltage regulation at each printed circuit board will be vastly improved and other power-distribution problems will be minimized. One of the minimized problems is noise or signal pickup on the long power-supply lines to remote locations.

IC voltage regulators are available in many types. Without referring to any specific type, we can say that internally they are considerably more complex than the CMOS amplifier studied previously. An

Voltage Regulators

LM340 series 3-terminal positive regulators

general description

The LM340-XX series of three terminal regulators is available with several fixed output voltages making them useful in a wide range of applications. One of these is local on card regulation, eliminating the distribution problems associated with single point regulation. The voltages available allow these regulators to be used in logic systems, instrumentation, HiFi, and other solid state electronic equipment. Although designed primarily as fixed voltage regulators these devices can be used with external components to obtain adjustable voltages and currents.

The LM340-XX series is available in two power packages. Both the plastic TO-220 and metal TO-3 packages allow these regulators to deliver over 1.0A if adequate heat sinking is provided. Even with over 1.0A of output current available the regulators are essentially blow-out proof. Current limiting is included to limit the peak output current to a safe value. Safe area protection for the output transistor is provided to limit internal power dissipation. If internal power dissipation becomes too high for the heat sinking provided, the thermal shutdown circuit takes over preventing the IC from overheating.

Considerable effort was expended to make the LM340-XX series of regulators easy to use and minimize the number of external components. It is not necessary to bypass the output, although this does improve transient response. Input by-passing is needed only if the regulator is located far from the filter capacitor of the power supply.

features

- Output current in excess of 1A
- Internal thermal overload protection
- No external components required
- Output transistor safe area protection
- Internal short circuit current limit
- Available in plastic TO-220 and metal TO-3 packages

voltage range

LM340-5	5V	LM340-15	15V
LM340-6	6V	LM340-18	18V
LM340-8	8V	LM340-24	24V
LM340-12	12V		

schematic and connection diagrams

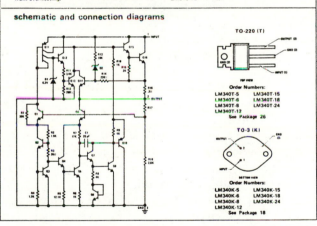

electrical characteristics (con't)

LM340-8 (V_{IN} = 14V, I_{OUT} = 500 mA, $0°C \leq T_A \leq 70°C$, unless otherwise specified)

PARAMETER	CONDITIONS	MIN	TYP	MAX	UNITS
Output Voltage	T_j = 25°C	7.7	8	8.3	V
Line Regulation	T_j = 25°C, 10.5V $\leq V_{IN} \leq$ 25V				
	I_{OUT} = 100 mA			80	mV
	I_{OUT} = 500 mA			160	mV
Load Regulation	T_j = 25°C, 5 mA $\leq I_{OUT} \leq$ 1.5A			160	mV
Output Voltage	10.5V $\leq V_{IN} \leq$ 23V, 5 mA $\leq I_{OUT} \leq$ 1.0A	7.6		8.4	V
	$P_D \leq$ 15W				
Quiescent Current	T_j = 25°C		7	10	mA
Quiescent Current Change	10.5V $\leq V_{IN} \leq$ 25V			1	mA
	5 mA $\leq I_{OUT} \leq$ 1.5A			.5	mA
Output Noise Voltage	T_A = 25°C, 10 Hz $\leq$ f $\leq$ 100 kHz		52		µV
Long Term Stability				32	mV/1000 hr
Ripple Rejection	I_{OUT} = 20 mA, f = 120 Hz		55		dB
Dropout Voltage	T_j = 25°C, I_{OUT} = 1.0A		2		V

Figure 8.31 Partial data sheet for an IC voltage-regulator circuit. (Courtesy of National Semiconductor.)

IC voltage regulator may have 50 internally interconnected devices, but we have treated the totality as a single component. Some regulators can be programmed by the user and may require one or more external components to determine the regulating voltage. Others regulate at fixed values and require no external components. Figures 8.30 and 8.31 show partial data sheets of presently available IC voltage regulators.

8.8   INTEGRATED-CIRCUIT DIFFERENTIAL AMPLIFIERS

Differential amplifiers have unique characteristics. They have two signal-input points and either one or two signal-output points. The output signal is an amplified form of the *difference* between the two instantaneous input-voltage signals. We will show a possible schematic diagram a little later and analyze the operation so that we can develop an intuitive feel for the concept of amplifying the difference voltage between two signals.

Differential amplifiers have been known, studied, and used for many years. They can be built using FETs, BJTs, or vacuum tubes. There are several situations in which it is desirable to amplify the difference voltage between two signals, and differential amplifiers are commonly used in such situations with near-perfect results.

Differential amplifiers have not always been studied in a first course or a basic course in electronics. However, when linear integrated circuits came into widespread use, the differential amplifier circuit drastically increased in importance. It is often used as a basic building block in complex IC circuits.

We are studying the differential amplifier for two reasons. First, to understand complex circuits containing differential amplifiers as internal components, it is very helpful to have a basic understanding of differential amplifiers. Second, some of the more complex circuits also offer the difference-of-two-signals input feature. But neither of these reasons is sufficient cause for us to do a detailed study of differential amplifier circuits. In fact, a detailed study would be an entire book and a complete course of study.

Please note that the title of this section is IC differential amplifiers. The IC portion is emphasized for three reasons. First, this section will help the reader understand the operation of the differential amplifiers in an IC manufacturer's product, but will not make him or her proficient at designing differential amplifiers using discrete transistors or tubes. Second, IC differential amplifiers are much more popular than are the discrete versions. Third, there are significant advantages of having a differential amplifier on a monolithic IC chip. On a monolithic chip, you may get a closer match of transistor parameters, and will for sure get better thermal tracking of the transistors. Thermal tracking means that if one transistor undergoes a temperature change, the other transistor will see an identical temperature change. Thermal tracking is important because transistor parameters change with temperature. If matched transistors in a differential amplifier connection undergo identical temperature changes, then ideally the output signal(s) will be unchanged by the temperature change. This sentence will make more sense after a study of the action of a differential amplifier stage and its inherent common-mode rejecting qualities.

A simple differential amplifier circuit is shown in Figure 8.32.

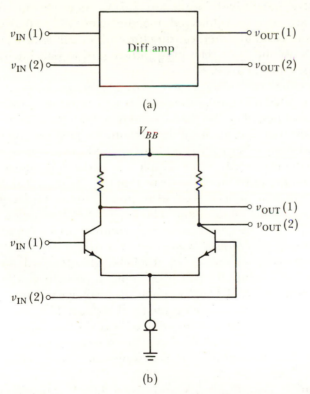

(a)

(b)

Figure 8.32 Simple differential amplifier: (a) block diagram; (b) schematic diagram.

Note that it consists of very few components—a field-effect diode (or any other type of constant-current source), two transistors, and two resistors. This circuit has differential outputs as well as differential inputs. Two input terminals and two output terminals are thus needed. The labeled voltages are, of course, with reference to the common, or ground, symbol.

Let us now analyze the response of a differential amplifier to various input voltages. In Figure 8.33 we see the results of such an analysis. The analysis requires little more than Ohm's law. Note that regardless of the input voltages, the sum of the emitter currents is 1 mA. Assume that the sum of the collector currents is also 1 mA. In Figure 8.33(a) the two transistor bases are at the same potential and the two emitters are at the same potential. This ensures that if the transistors are identical, the emitter currents will be equal. With 0.5 mA of current through each collector 10-kΩ resistor, there is a 5-V drop across each one. The result is that both V_{C1} and V_{C2} are at 10 V wrt ground.

In Figure 8.33(b) both inputs are at 7 V. As in part (a), the two emitters are directly connected together and the two bases are directly connected together, causing the emitter currents to be equal at 0.5 mA. By the same reasoning with regard to part (a), V_{C1} and V_{C2} will be equal at 10 V.

The sample input voltages for Figure 8.33(a) and (b) were chosen to illustrate what is known as the *common-mode* rejecting properties of a differential amplifier. By our definition, common mode refers to two signals that are identical in amplitude, polarity, frequency, phase, and every other respect and are applied to the two inputs of a differential amplifier. In our example, we had a 2-V change in the common-mode signal inputs and saw that the output-signal voltages did not change at all. This is an ideal situation that cannot quite be obtained in a practical situation. The measure of how well a differential amplifier can ignore or reject common-mode input-signal voltages is called its *common mode rejection* (CMR). CMR is measured in units of volts per volt or in decibels (dB). (Decibel units are considered in Chapter 9.) If, for example, a particular differential amplifier has a CMR of 1 mV/V, the output voltage would change 1 mV for a 1 V common-mode input-signal voltage. This corresponds to CMR = 60 dB. CMR, as defined here, is really the common-mode voltage gain of the circuit. However, since the common-mode gain is much less than 1, the word rejection sometimes seems more appropriate than gain.

In Figure 8.33(c), $V_{B1} = 5$ V and $V_{B2} = 4.95$ V. The emitters are tied together, so they therefore must be at the same potential. The preceding facts make it apparent that V_{BE} of Q_1 is greater than V_{BE} of Q_2 by 50 mV. Transistor Q_1 will therefore have more emitter (and collector) current than will Q_2. Our assumed current division shows I_{E1} at 0.8 mA and I_{E2} at 0.2 mA. Completing the circuit analysis, we find that $V_{C1} = 7$ V and $V_{C2} = 13$ V.

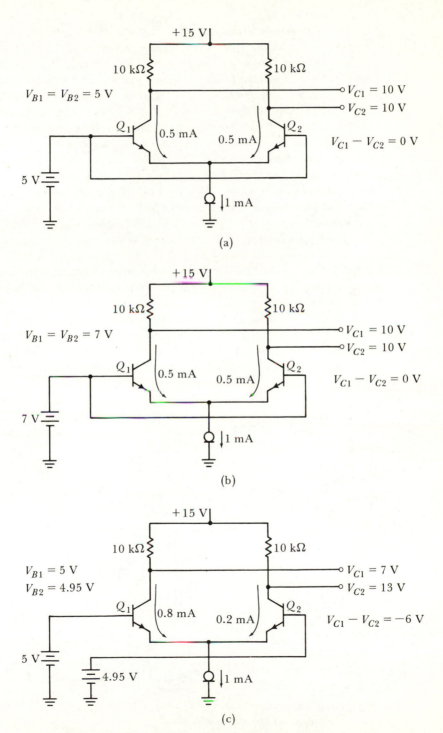

Figure 8.33 Sample input-output voltages for differential amplifiers.

You may wonder why we chose an example with only 0.05 V between input leads. We did so because differential amplifiers normally have very large gains. At least, differential amplifiers made with BJTs have relatively large voltage gains. If the input difference voltage for a circuit such as that of Figure 8.32 becomes as large as 0.2 or 0.3 V, one of the transistors would be completely off and we will no longer have an operating differential amplifier circuit.

We use the word *amplifier* as part of the descriptive name of the differential amplifier circuit. Therefore we need to look at the amplifying or gain properties of the circuit. There are two distinct cases we need to consider. These cases are the single-ended-output case and the differential-ended-output case. For our example circuits shown in Figure 8.33, the output is single-ended if the voltage V_{C1} (or V_{C2}) wrt ground is considered as the output-signal voltage. The output is differential-ended if the difference between V_{C1} and V_{C2} is considered as the output-signal voltage. Both cases are important. There are a significant number of applications for both the differential-ended and the single-ended cases.

The simple circuit we have used for explanation purposes used only one voltage polarity. In many cases, however, differential amplifier circuits are intended to be operated from two opposite-polarity voltage sources. Sources of ± 15 V are common for the IC variety differential amplifier. The dual-supply concept is shown in Figure 8.34. The input- (and output-) signal voltages are usually referenced to ground in this situation, which means that no dc biasing of the input leads is necessary. We will take advantage of this,

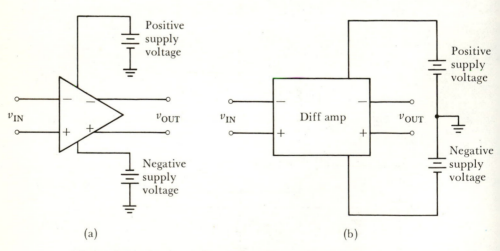

Figure 8.34 Dual-supply-voltage differential amplifier connections: (a) triangular symbol; (b) block symbol.

as you may note, in following figures where signal sources are referenced to ground. In Figure 8.34, we have shown alternate symbols. It is common to use the differential amplifier symbols without the supply-voltage terminals shown. The supply voltages are then assumed to be correctly connected.

Now let us give some thought to how we would measure the amplification or voltage-gain properties of a differential amplifier. In Figure 8.35 we see several connections for measuring the common-mode differential-ended voltage gain $A_V(CMDE)$.

In Figure 8.35(a) we have the preferred connection for measuring $A_V(CMDE)$. Note that the inputs are tied together so that the signal input must be common mode. Also note that the signal input is referenced to ground. No bias circuits are needed. To get data, merely set up the signal generator for some moderate frequency and moderate amplitude. Measure and record the peak-to-peak signals in and out of the differential amplifier. Then calculate the common-mode differential-ended voltage gain as

$$A_V(CMDE) = \frac{V_{out}}{V_{in}}$$

In Figure 8.35(b) we have the all-too-common situation where one of the oscilloscope input-signal leads is internally tied to ground. Do not use such an oscilloscope as indicated. We will consider alternate ways to get the desired data.

If your oscilloscope with one signal lead grounded is a dual-trace unit capable of operating in differential mode, then the connection shown in Figure 8.35(c) will work fine. It is quite similar to the situation in (a).

If you must use a single-trace oscilloscope with one signal-input line grounded, the connection of Figure 8.35(d) is available. With the input-voltage source set at some known level, measure and record both $V_{OUT}(1)$ and $V_{OUT}(2)$. The differential output is of course $V_{OUT}(1) - V_{OUT}(2)$, or vice versa. Now do the same with another known input voltage. Calculate the common-mode differential-ended voltage gain as

$A_V(CMDE)$
$$= \frac{\Delta V_{OUT}}{\Delta V_{IN}} = \frac{[V_{OUT}(1) - V_{OUT}(2)]_1 - [V_{OUT}(1) - V_{OUT}(2)]_2}{(V_{IN})_1 - (V_{IN})_2}$$

The subscripts 1 and 2 refer to the first and second chosen input voltages.

If you should want the common-mode gain with a single-ended output, you can merely use the connection as in Figure 8.35(a)

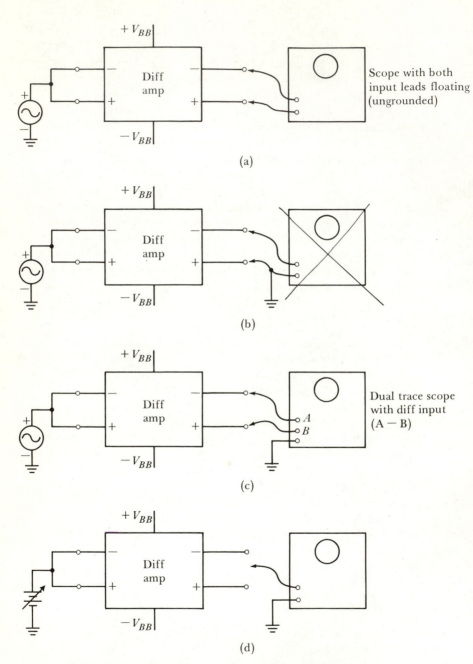

Figure 8.35   Alternate circuit connections for measuring common-mode voltage gain: (a), (c), (d) correct methods; (b) wrong method.

except that you must connect the oscilloscope from one signal-output point to ground. Calculate the common-mode single-ended voltage gain as

$$A_V(\text{CMSE}) = \frac{V_{\text{out}}(\text{SE})}{V_i}$$

The SE notation associated with the output voltage is to remind us that the output voltage is the single-ended voltage measured from one output terminal wrt ground.

To measure the differential-mode voltage gain of a differential amplifier, we use circuit connections as in Figure 8.36. In (a) and (b) we have assumed the use of an oscilloscope with both input points

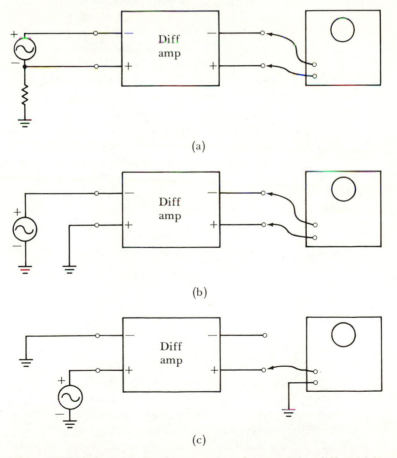

(a)

(b)

(c)

Figure 8.36   Alternate circuit connections for measuring differential-mode voltage gain: (a), (b) differential output; (c) single-ended output.

isolated from ground. If such an oscilloscope is not available, you must use some of the techniques shown previously in Figure 8.35.

Note the signal-input circuits shown in Figure 8.36. The resistor shown in (a) is sometimes needed. Its purpose is to establish a reference between the signal source and the differential amplifier. Sometimes the necessary reference is provided through the input impedance of the differential amplifier. If the input impedance of the amplifier is extremely high (as it is if FETs are used on the signal-input lines of the amplifier), and if no reference path is provided, then the signal source may pick up a static charge such that its signal voltage is riding on a large dc level wrt the amplifier reference. It is true that a dc level is a common-mode signal that a differential amplifier should ignore. However, there are limits to the CMR properties of an amplifier. To convince yourself of a limit to the permissible common-mode voltage level, go back to Figure 8.32 and try to explain proper operation of the circuit if there were a 1-kV common-mode voltage. The value of the reference resistor is not critical. In fact, it can be zero as in Figure 8.36(b) and (c). The signal-input circuits shown in Figure 8.36 are the same, except as follows. In (a) the reference resistor is a nonzero value. In (b), the reference resistor is zero. Part (c) is the same as (a), except that the opposite input terminal is referenced to ground.

The measurements and calculations for determining the differential-mode gain are quite similar to that already done for the common-mode case. Just use an oscilloscope, or some other suitable voltage-measuring device, to observe the input- and output-signal-voltage levels. Calculate the gain using one of the following equations.

$$A_V(\text{DMDE}) = \frac{V_{\text{out}}(\text{DE})}{V_{\text{in}}}$$

$$A_V(\text{DMSE}) = \frac{V_{\text{out}}(\text{SE})}{V_{\text{in}}}$$

The only difference in the equations has to do with whether we consider the output from one output terminal to ground (SE case) or whether we consider the output signal to be the voltage between the two output terminals (DE case). An oscilloscope is the preferred instrument for this gain measurement (in the author's opinion), since the waveshape is visible and distortion will be readily apparent. Because differential amplifiers normally have high gain, we must be careful with the amplitude of the input voltage so that the output will not be distorted. In some cases, just a few millivolts of differential input signal may be enough to cause the internal amplifiers to

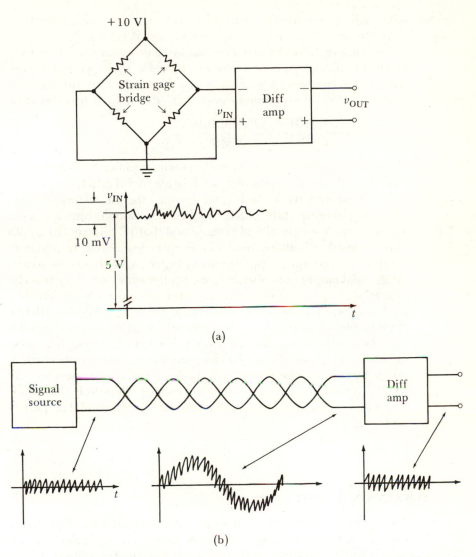

Figure 8.37   Differential amplifier applications: (a) strain-gage-bridge and amplifier application; (b) common-mode noise application. (The "strain gage bridge notation refers to the four-resistor diamond-shaped network rather than to one resistor as implied in the drawing.)

slam against the usable limits of their respective load lines with resulting severe distortion of the output-signal voltage.

Now that we have discussed the various gain parameters of differential amplifiers, we are in a position to define a new term. The new term is the *common-mode rejection-ratio* (CMRR). It is the ratio of differential-mode to common-mode gain. In equation form we have

$$\text{CMRR} = \frac{A_V(\text{DMDE})}{A_V(\text{CMDE})} = \frac{A_V(\text{DMSE})}{A_V(\text{CMSE})}$$

Since we would like to have a differential amplifier with zero common-mode gain, it follows that we desire a high CMRR.

Now that we have looked at the circuits, the characteristics, and the terminology of differential amplifiers, we should mention some applications. We have already mentioned that IC differential amplifiers are used as building blocks in complex linear ICs. In addition, we show two common applications in Figure 8.37. In (a) we have a differential amplifier as a strain-gage-bridge amplifier. Note that the desired signal of just a few millivolts is riding on a 5-V level. The 5-V level is a common-mode signal and is therefore ignored (mostly) by the differential amplifier. In (b) we have a situation where a signal is sent some significant distance over twisted wires through a noisy environment. Noise of a 60-Hz sine wave form is commonly encountered. Since the noise is common mode, it will appear only in very small proportions in the output-signal voltage. Hopefully the reader now has a reasonably good understanding of differential amplifiers and their uses.

8.9   OPERATIONAL AMPLIFIERS

The operational amplifier (op amp) is a widely used product that is readily available in integrated-circuit form. The op amp may be quite complex internally. Items such as differential amplifiers, FETs, BJTs, Zener diodes, constant-current sources, resistors, and *PN*-junction diodes are commonly found in the internal makeup of operational amplifiers. As we have done previously, we will consider the entire op amp as a component.

The operational amplifier derives its name from the fact that it can readily perform mathematical operations. We will show how the op amp can perform the mathematical operations of addition, subtraction, multiplication, division, integration, and differentiation.

An idealized, or perfect, op amp has a distinct set of terminal characteristics. It must be dc-coupled so that it can operate on steady-state or dc values. Additionally, the dc levels must be such that zero voltage at the input terminal results in zero voltage at the output terminal. Any deviation from this condition is called an offset voltage and is undesirable. The op amp must have a polarity inversion. In addition to these conditions, an ideal op amp is characterized by the following numerical values.

$A_V = -\infty$ (minus sign signifies polarity inversion)
$Z_{in} = \infty$
$Z_{out} = 0$

Ideal components, of course, exist only in our imagination. You may be surprised, however, to discover some of the characteristics which are attainable for IC op amps. It is not particularly difficult, for example, to obtain op amps with characteristics such as

$A_V = -60{,}000$

$Z_{in} = 10^{12} \ \Omega$

$Z_{out} = 2 \ \Omega$

Hopefully, this is sufficient to convince you that available IC op amps will behave very much like the ideal op amps that we will be discussing.

Integrated-circuit op amps are very reasonably priced. It is not unusual to find that a capacitor or a power resistor will cost as much as an op amp. Op amps may be purchased with one, two, or four units in one package. Several package configurations are available.

Figure 8.38 gives block-diagram representations of op amps. Note that both inverting and noninverting inputs are shown. This is an indication of the fact that one or more differential amplifier is used internally to build the op amp. When we speak of the voltage gain of an op amp, we are actually speaking of the differential-mode single-ended voltage gain. In many cases only the inverting input will be

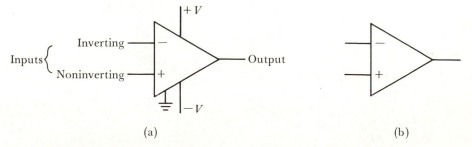

(a)   (b)

Figure 8.38 Block diagrams of op amps: (a) with labeled terminals and showing supply voltages; (b) normal representation.

used. For simplicity we will not normally show the power-supply connections. Thus we have a two- or three-terminal device for study and consideration.

An op amp is normally used with at least two external components. The circuit connection with the two additional components is shown in Figure 8.39. Note that the input to the circuit is different from either of the input terminals of the op-amp component. Signals at the input and output terminals are assumed to be with respect to some common system reference, probably the same point to which the noninverting input is connected. In cases where only one input terminal and one output terminal is shown, as in Figure 8.39, the other terminal is implied to be the system common-reference point.

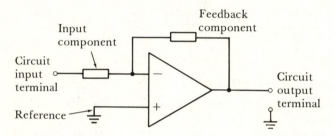

Figure 8.39   Operational-amplifier circuit.

Let us carefully consider the circuit of Figure 8.40 to gain an understanding of how an op-amp circuit operates. Note that the inverting input terminal on the op amp is called virtual ground. This is a consequence of the component's having infinite (ideal) gain. If the inverting input terminal should be other than at zero volts, this input voltage would be amplified by infinity, giving the circuit an infinite output voltage. By our previous study of BJTs and FETs, we

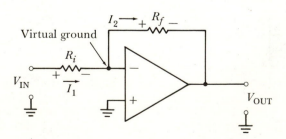

Figure 8.40   Op-amp circuit used to explain operation.

know that the dc output voltage of an amplifier cannot exceed the supply voltage. Since the op-amp supply voltage is definitely not near infinity (± 15 V is typical), the output-terminal voltage cannot be a large value. If the output voltage must be in the ± 15 V-range, and by the gain relationship we know that the output voltage is infinitely larger in magnitude than the input-terminal voltage, then we must conclude that the inverting input-terminal voltage must be infinitely small or zero. Since the inverting input terminal must remain at a near-zero voltage level, the virtual-ground terminology is seen to be appropriate.

We have just determined that the voltage to the inverting input terminal must remain zero (or very nearly so). Let us refer again to Figure 8.40 to see if we can get an intuitive feel for why a circuit connected in this way will always maintain the inverting input terminal at zero voltage. Assume that a positive V_{IN} voltage is applied. Also assume that the inverting input terminal tends to follow the V_{IN} voltage in the positive direction. As the inverting input terminal goes just slightly positive, the output voltage must go negative owing to the inverting quality of the op amp. Under the conditions just assumed, the following circumstances should be noted. We have two resistors connected to the inverting input terminal. One of these resistors (R_i) goes to the positive input voltage V_{IN} and the other resistor (R_f) goes to the output-terminal voltage V_{OUT}. Since the inverting input terminal is a junction of a voltage-divider network consisting of R_i and R_f and the resistors connect to opposite polarity voltages, it is possible that the junction voltage (that is, the inverting-input terminal) might be at zero voltage.

We have stated in the preceding paragraph that the voltage at the inverting input terminal might be zero. Now let us go further and show that the terminal must be at zero voltage. We will assume nonzero conditions and then show that nonzero conditions cannot be maintained. First let us assume that the inverting input terminal tends to follow the input voltage V_{IN} and takes on a significant positive voltage. Owing to the large inverting gain, this positive voltage on the inverting input terminal will result in a much larger negative voltage on the output terminal. This negative voltage is on one end of the voltage divider consisting of R_f and R_i so it will tend to move the voltage-divider junction (that is, the inverting input terminal) back toward the zero voltage from which we assumed it had departed.

Now let us assume that the inverting input terminal tries to assume a significant negative value. If it does, then because of the large inverting gain, the output terminal must take on a much larger positive value. Our assumptions have put us in the assumed position of having both ends of the dc voltage-divider network at a positive voltage and the junction at a negative voltage. This is obviously an unstable situation. If the assumed condition should somehow become true at some instant of time, the junction (that is, the inverting input terminal) would definitely move back toward the zero-voltage point. Our assumption of the inverting input terminal leaving zero in the negative direction was thus a false assumption.

We have just shown that the inverting input cannot take on either a positive or negative value. Thus it has no choice but to remain at zero voltage. Naming this point a virtual ground seems quite appropriate.

The preceding discussion was based upon an op amp with ideal characteristics. Just this one time, let us assume a nonideal condition to see if the conclusion about the inverting input terminal being a virtual ground is valid. Assume that the voltage gain of the op amp is 50,000 instead of an ideal value of infinity. As before, if the inverting input terminal tends to stray from zero, the output-terminal voltage will go in the opposite direction, which will tend to return the inverting input terminal toward ground. In this case, the output terminal will go 50,000 times as far as the inverting input terminal in an effort to return the input toward ground. In a typical case, with ±15-V power supplies and maximum permissible signal levels, the output terminal could swing nearly 15 V in either direction. The inverting input terminal voltage under this condition (that is, one that causes the ±15 V on the output terminal) can be calculated as 0.0003 V (0.3 mV) using the known gain relationship. Thus we see that the inverting input terminal does not stay at exactly ground potential. It is very close, however. We will henceforth call the inverting input terminal, quite appropriately, a virtual ground when the op amp is connected as in Figure 8.40.

We are now ready to see how an op-amp circuit can do mathematical operations. Again we refer to Figure 8.40. The current arrows and voltage-drop polarity signs are appropriate for the case where V_{IN} is a positive voltage. Based on the virtual-ground concept, the voltage across R_i is V_{IN} volts. This is the case because one end of R_i is connected to V_{IN} and the other end is tied to (virtual) ground. The current through R_i is

$$I_1 = \frac{V_{IN}}{R_i}$$

Since an op amp has infinite input resistance, none of the I_1 current may flow into the inverting input terminal. Of necessity, then, all of I_1 must flow through R_f. This requires that

$$I_1 = I_2$$

We now know the current through R_f, so we can calculate the voltage across R_f.

$$V_{Rf} = I_2 R_f = I_1 R_f$$

Since we have already solved for the value of I_1, we will substitute this value into the equation for R_f voltage.

$$V_{Rf} = I_1 R_f = \frac{V_{IN}}{R_i} R_f$$

Now let us examine the voltage across R_f. Note that R_f is connected from the output terminal to virtual ground. Also note that the output voltage V_{OUT} is taken between output terminal and actual ground. Thus we should be able to convince ourselves that the voltage across R_f and the output voltage are identical.

$$V_{OUT} = V_{Rf} = \frac{V_{IN} R_f}{R_i}$$

If we solve this equation for the voltage gain between V_{IN} and V_{OUT} terminals, we have

$$A_V = \frac{V_{OUT}}{V_{IN}} = \frac{R_f}{R_i}$$

To be correct in a strict mathematical sense, we should note that the output terminal connects to the more negative end of the R_f voltage. Thus

$$A_V = \frac{V_{OUT}}{V_{IN}} = -\frac{R_f}{R_i}$$

The minus sign signifies a polarity inversion. Even though the derivation of gain has been done with assumed dc quantities, it is valid for any input.

The voltage-gain equation for the standard op-amp-circuit configuration shown in Figure 8.40 is quite interesting. Note that the gain is totally a function of the two resistors. Any voltage gain is available to the designer merely by picking appropriate resistance values. Even for a fixed gain value, the absolute resistance values are not fixed. Only the ratio of resistance is fixed. Gain can theoretically take on any value between zero and infinity.

Remember that there is a polarity inversion associated with a standard op-amp-circuit configuration. If a noninverting gain is needed, you can use another inverting op amp having unity gain. The two-stage amplifier will have the desired gain and a 360° (equivalent to 0°) phase shift. You can also proportion the gain between stages if you remember that the overall gain will be the product of the individual stage gains.

Now let us consider another op-amp-circuit configuration. The circuit is shown in Figure 8.41. Since the inverting input has infinite input resistance, and thus cannot draw any current, we know that

$$I_5 = I_1 + I_2 + I_3 + I_4$$

Knowing the current through R_f, we can calculate the voltage drop across R_f.

$$V_{Rf} = I_5 R_f = (I_1 + I_2 + I_3 + I_4)R_f$$

As before, we note that the output voltage V_{OUT} is the negative of the voltage across R_f.

$$V_{OUT} = -(I_1 + I_2 + I_3 + I_4)R_f$$

Making use of the virtual-ground phenomenon, we can easily calculate the current values I_1, I_2, I_3, and I_4 and substitute these values into the V_{OUT} equation. Doing so we obtain

$$V_{OUT} = -\left(\frac{V_1}{R_1} + \frac{V_2}{R_2} + \frac{V_3}{R_3} + \frac{V_4}{R_4}\right)R_f$$

If all the input resistors are of equal value, the equation simplifies to

$$V_{OUT} = -\frac{R_f}{R_1}(V_1 + V_2 + V_3 + V_4)$$

When the equation is written in this form, it is quite easy to see why the op amp is said to be able to perform the arithmetic operation of addition. If the negative sign is objectionable, follow the summing stage with a unity-gain inverting op-amp circuit.

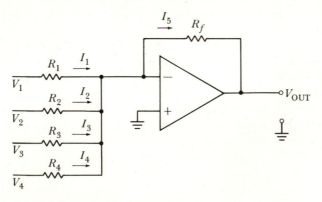

Figure 8.41 Op-amp summing circuit.

If a combined addition and subtraction operation is required, it may be performed with two stages, as shown in Figure 8.42(a). Note that all R_i values are the same. The equation solved by this circuit is shown near the output terminal.

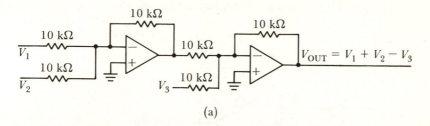

(a)

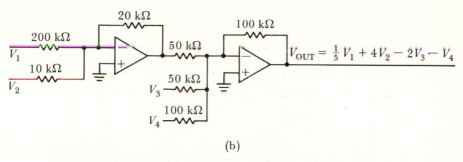

(b)

Figure 8.42 Op-amp arithmetic circuits: (a) combined addition and subtraction operation; (b) combined four-function operation.

Multiplication and division are performed by an op amp because of its gain characteristics. If a multiplication by 10 is desired, make the R_f/R_i ratio equal to 10. Then the result will be

$$V_{OUT} = -10V_{IN}$$

If a division by 10 is desired, make the R_f/R_i ratio equal to one-tenth. Then the result will be

$$V_{OUT} = -\frac{V_{IN}}{10}$$

In Figure 8.42 we have op-amp circuits connected to perform combined arithmetic operations. All resistors are shown with specific numerical values. The circuit of Figure 8.42(a) performs a combined addition and subtraction operation. The circuit of Figure 8.42(b) combines all four of the simple arithmetic operations. With some careful thinking, you should be able to verify the output equations as shown. In each case you can find the output of the first stage. This output becomes one of the inputs to the second stage. You should now be thoroughly convinced of the versatility and usefulness of op amps.

An op-amp circuit is very efficient at performing the mathematical operation of integration. In Figure 8.43 we have a schematic drawing of an integrating circuit. Remembering the virtual ground at the

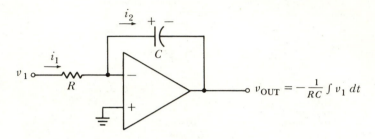

Figure 8.43   Op-amp integrating circuit.

inverting input terminal, we can solve for the current i_1.

$$i_1 = \frac{v_1}{R}$$

Since the currents i_1 and i_2 are equal, we know that

$$i_1 = i_2 = \frac{v_1}{R}$$

Now that we know the current through capacitor C, we want to find the voltage across C. Recalling earlier studies in circuit theory, we know that the voltage-current relationship for capacitors may be written as

$$i_C = C \frac{dv_C}{dt}$$

or

$$v_C = \frac{1}{C} \int i_C \, dt$$

(It should be noted that in a strict mathematical sense the equation should be

$$V_C = \frac{1}{C} \int i_C \, dt + \text{constant}$$

where the constant represents the initial terminal voltage of the capacitor.) In this case we are looking for the voltage across the capacitor. Noting that v_{out} is the negative of the voltage across C, and substituting for the current, we obtain

$$v_{OUT} = -v_C = -\frac{1}{C}\int \frac{v_1}{R}\, dt \text{ V}$$

Since R is normally a constant, we can bring it out in front of the integral sign. In a most common form we thus have

$$v_{OUT} = -\frac{1}{RC}\int v_1\, dt \text{ V}$$

The output voltage is thus seen to be the integral of the input voltage multiplied by a constant value $(-1/RC)$.

There are many applications for integrating circuits. Three applications come to mind immediately.

1. Analog-computer circuits.

2. Waveshaping circuits.

3. Phase-shifting circuits.

Integrating circuits make up a large percentage of all the active stages in most analog-computer applications. A common waveshaping example is the case where a square-wave input signal is integrated to form a triangle-wave output signal. A phase shift of 90° (actually $-90°$) occurs when a sinusoidal input waveform is applied to an integrating circuit. The output waveform will be sinusoidal in shape but shifted in phase by 90°. For example, assume that the input waveform is

$$v_1 = 10 \sin \omega t \text{ V}$$

Then we have

$$v_{OUT} = -\frac{1}{RC}\int v_1\, dt = -\frac{1}{RC}\int 10 \sin \omega t\, dt \text{ V}$$

Our previous mathematical studies should have made us aware that

$$\int 10 \sin \omega t\, dt = -\frac{10}{\omega}\cos \omega t$$

Thus we have

$$v_{OUT} = \frac{10}{\omega RC}\cos \omega t \text{ V}$$

If you plot waveforms for the input (sin) and output (cos) signals, you will find that they differ in phase by exactly 90°.

An op-amp differentiating circuit is shown in Figure 8.44. The

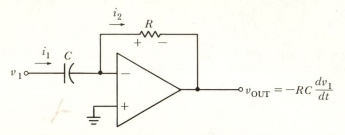

Figure 8.44 Op-amp differentiating circuit.

analysis of this circuit proceeds as in previous examples. The current i_1 is the current through capacitance C.

$$i_1 = C\,\frac{dv_1}{dt} = i_2$$

The output voltage is the negative of the voltage across R.

$$v_{\text{OUT}} = -RC\,\frac{dv_1}{dt}$$

The differentiator circuit is shown for completeness. It is a seldom-used circuit. It is very susceptible to noise. Whereas all circuits respond to noise in some way, the differentiator circuit responds more favorably to most noise signals than it does to the signals we are trying to process.

Noninverting amplifiers are possible with differential-input op amps. In Figure 8.45 we have a so-called voltage-follower circuit. It is in reality a noninverting unity-gain amplifier circuit.

When using the differential-input feature of op-amp circuits, the infinite-gain feature is considered somewhat differently. In this case

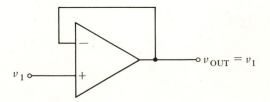

Figure 8.45 Op-amp voltage-follower circuit.

the output-terminal voltage, with respect to ground, is (ideally) infinitely larger than the voltage difference between the two input terminals. Thus the only way the output voltage can stay within bounds, typically ±15 V for IC versions, is for the voltage difference between the input terminals to be infinitely small or zero. Applying this to the practical case, we always assume that the two input terminals are at the same voltage, that is, there is no difference in voltage.

Let us now return to the voltage-follower circuit of Figure 8.45. Assume that the input voltage v_1 begins to move from zero, wrt ground, in the positive direction. Since the input is connected to the noninverting input, the output voltage will also move in the positive direction. Since the output terminal is directly connected to the inverting input terminal, the output must move exactly the same amount as the input so that the voltage difference between the input terminals remains zero. When the input goes negative, a similar circumstance occurs. Thus the output and input voltages are always the same. This circuit is in fact a noninverting unity-gain amplifier.

Since the voltage follower has the same input and output waveforms, you may wonder about its usefulness. One application of voltage followers is to obtain isolation, that is, to reduce the interaction between circuits. When used for this purpose, an amplifier is called a *buffer amplifier*. Another application concerns the case where a voltage follower is used for its low output-impedance property.

The voltage-follower principle can be modified in order to obtain a noninverting amplifier with gain. Such a circuit is shown in Figure 8.46. In this case R_1 and R_2 form a voltage-divider network so that

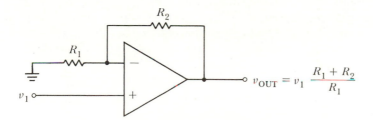

Figure 8.46   Op-amp noninverting amplifier with gain.

only a portion of the output-signal voltage is returned to the inverting input terminal. Assume, for example, that resistors R_1 and R_2 are chosen such that one-fourth of the output voltage appears at the inverting input terminal. Then in order for the inverting input to be at the same voltage as the noninverting input terminal, that is, for the voltage difference between the input terminals to be zero, the output voltage must be four times the amplitude of the input voltage. This, of course, describes an amplifier circuit.

This concludes our study of op-amp circuits. By this time you should have a good solid understanding of the basic operating principles of the op amp as a component. You are not an expert yet, however. There seems to be an unlimited number of applications for op amps. If you are interested in further study of the op amp and its applications, you should refer to the books written by the staff of Burr-Brown Research Corporation.

8.10   IC INSTRUMENTATION AMPLIFIERS

Instrumentation amplifiers consist of a unique and dedicated interconnection of operational amplifiers. The net result is that an IC instrumentation amplifier is a high-quality amplifier with differential inputs and a single-ended output. Normally users have control over the operation of the instrumentation amplifier only in the sense that they can control the differential-mode single-ended voltage gain of the device. This gain-control function is made available to the operator in the form of either one or two resistors that are connected externally to the IC package. The gain is specified by the manufacturer as a known function of the external resistor(s).

Maybe we should pause at this point to emphasize that an IC instrumentation amplifier is distinctly different from an IC operational amplifier. As we have seen in our previous studies, an operational amplifier is capable of performing various mathematical operations on an input signal. The function to be performed is dependent on the external components used. By contrast, an instrumentation amplifier is capable of performing only an amplifying function.

Ideally, an instrumentation amplifier has no dc offset voltage, which means that if there are no signal voltages on the input terminals, or if there is only a common-mode signal on the input terminals, the output voltage will be zero. Since the ideal situation cannot be totally achieved in a practical situation, the IC instrumentation amplifier usually has some provision for adjusting the offset to zero by the use of external components. The user will need to consult the manufacturer's data sheet for details on making offset adjustments on any specific type of IC instrumentation amplifier.

In Figure 8.47 we see how an instrumentation amplifier is made up of three operational amplifiers. The dashed-line enclosure indicates components that are contained within the IC package. Note

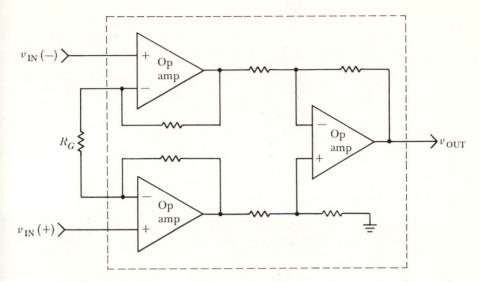

Figure 8.47 Block-diagram representation of an IC instrumentation amplifier.

that only the gain-determining resistor R_G is external and therefore available to the user. (Some IC instrumentation amplifiers have two external gain-determining resistors.) The two input terminals identified by the polarity signs in parentheses indicate that the instrumentation amplifier has differential inputs. As is common in similar circumstances, the supply-voltage terminals are not shown. Dual supply voltages are commonly specified with ±15 V being a typical value.

A triangular symbol is commonly used to indicate an amplifier of any type. An instrumentation amplifier is no exception. In Figure 8.48 we have possible block symbols of both triangular and rectangular form. In many cases the supply-voltage terminals and the ground-reference terminal (if any) will not be shown.

A circuit analysis of an instrumentation amplifier may appear to be a formidable task. However, if we take advantage of previous learning about operational amplifiers and approach the problem in a logical and appropriate manner, the analysis is not difficult.

In Figure 8.49 we have a block representation of an instrumentation amplifier with all passive components labeled. Several points of interest, the voltages of which we are interested in, are labeled with a subscripted V. These voltages are with respect to ground. The operational amplifiers are assumed to be ideal.

Basically, our analysis boils down to a repeated application of Ohm's law and of making use of a fact that we learned about when studying differential-input operational amplifiers. The important

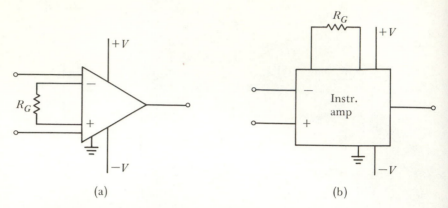

Figure 8.48 Instrumentation amplifier block-diagram representation: (a) triangular form; (b) rectangular form.

fact can be stated as follows: The input-terminal voltages must be equal for an ideal differential-input op amp whose output-terminal voltage is in a normal operating region. This is owing to the infinite gain of an ideal op amp. Any differential input at all would receive infinite gain and would cause the output-terminal voltage to try to go to infinity. Remember that the input-terminal voltages must be equal for each individual op amp.

Let us now discuss the circuit of Figure 8.49. We will use the indicated input voltages of $V_1 = 10$ V and $V_2 = 0$ V. Thus the differential input is 10 V. We will calculate several internal voltages and currents until we arrive at the output-terminal voltage.

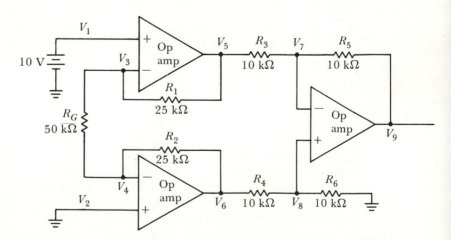

Figure 8.49 An instrumentation amplifier labeled for analysis purposes.

Since the input voltages V_1 and V_2 are given, we determine by inspection that

$$V_3 = V_1 = 10 \text{ V} \quad \text{and} \quad V_4 = V_2 = 0 \text{ V}$$

Again by inspection we determine that

$$|V_{RG}| = V_3 - V_4 = 10 \text{ V}$$

and

$$I_{RG} = \frac{V_3 - V_4}{R_G} = \frac{10 \text{ V}}{50 \text{ k}\Omega} = 0.2 \text{ mA} \downarrow$$

The arrow associated with the current I_{RG} indicates the direction of conventional current flow through the resistor R_G as configured in the diagram of Figure 8.49. Similar meanings are implied for arrows drawn up, to the left, or to the right.

Since we now know the magnitude and direction of I_{RG}, we can determine the magnitude and direction of I_{R1} and I_{R2}. The input terminals of the ideal op amps have infinite impedance and therefore will source or sink no current. The result, by inspection, is that

$$I_{R1} = \overset{\longleftarrow}{0.2 \text{ mA}} \quad \text{and} \quad I_{R2} = \overset{\longrightarrow}{0.2 \text{ mA}}$$

The voltage drops across resistors R_1 and R_2 can now be calculated.

$$V_{R1} = I_{R1}R_1 = (0.2 \text{ mA})(25 \text{ k}\Omega) = -(5 \text{ V}) +$$

$$V_{R2} = I_{R2}R_2 = (0.2 \text{ mA})(25 \text{ k}\Omega) = +(5 \text{ V}) -$$

The polarity signs associated with the calculated voltages indicate the relative polarity of the respective right and left ends of the resistors as configured in Figure 8.49.

Now that we know V_3, V_4, and the voltage drops across the resistors R_1 and R_2, we are in a good position to calculate V_5 and V_6.

$$V_5 = V_3 + V_{R1} = (10 + 5) \text{ V} = 15 \text{ V}$$

$$V_6 = V_4 - V_{R2} = (0 - 5) \text{ V} = -5 \text{ V}$$

Note that the magnitude of the voltage V_8 is determined by the voltage-divider resistors R_4 and R_6. Since the resistors R_4 and R_6 are equal, and since the input impedance of the op amp is infinite, the voltage V_8 is one-half the magnitude of V_6.

$$V_8 = V_6 \frac{R_6}{R_4 + R_6} = (\tfrac{1}{2})(5) = -2.5 \text{ V}$$

The voltage on the other input terminal of the op amp must be the same.

$$V_7 = V_8 = -2.5 \text{ V}$$

The current I_{R3} can be calculated based on the now known voltages V_5 and V_7.

$$I_{R3} = \frac{V_7 - V_5}{R_3} = \frac{(-2.5 \text{ V}) - (15 \text{ V})}{10 \text{ k}\Omega} = \frac{-17.5 \text{ V}}{10 \text{ k}\Omega} = \overrightarrow{1.75 \text{ mA}}$$

Owing to the infinite input impedance of the op amp, it should be apparent that

$$I_{R5} = I_{R3} = \overrightarrow{1.75 \text{ mA}}$$

Now we are (finally) in a position to calculate the voltage across R_5 and the output voltage V_9.

$$V_{R5} = I_{R5}R_5 = (1.75 \text{ mA})(10 \text{ k}\Omega) = +(17.5 \text{ V}) -$$

$$V_9 = V_7 - V_{R5} = (-2.5 \text{ V}) - (17.5 \text{ V}) = -20 \text{ V}$$

Now that we have calculated our way through the instrumentation amplifier, let us think about the results obtained. We had a differential-input voltage of 10 V and found that the resultant output-voltage magnitude was 20 V. The gain of the amplifier is

$$A_V = \frac{v_{\text{OUT}}}{v_{\text{IN}}} = \frac{20 \text{ V}}{10 \text{ V}} = 2$$

A similar analysis can be done for any permissible set of input voltages. The results obtained from our example analysis, and of three other sets of input values, are entered in Table 8.1. The first two examples were chosen to show that the gain remained at 2 as the inputs varied. Examples 3 and 4 were chosen to show that, in the ideal case, the instrumentation amplifier has infinite common-mode rejection and zero offset. The variables are listed (top to bottom) in Table 8.1 in the same order in which they should be calculated. Hopefully each student will do enough analysis of the circuit to verify the contents of the table.

The voltage gain for the instrumentation amplifier we have studied can be stated as

$$A_V = 1 + \frac{2R_1}{R_G}$$

able 8.1 A tabulation of calculated quantities for the instrumentation amplifier of Figure 49

	Method of calculating	Example 1	Example 2	Example 3	Example 4
1	Given	+10 V	0	0	−5 V
3	$= V_1$	+10 V	0	0	−5 V
2	Given	0	+5 V	0	−5 V
4	$= V_2$	0	+5 V	0	−5 V
RG	$\left\|\dfrac{V_3 - V_4}{R_G}\right\|$	↓ 0.2 mA	↑ 0.1 mA	0	0
R1	$= I_{RG}$	← 0.2 mA	→ 0.1 mA	0	0
R2	$= I_{RG}$	0.2 mA	0.1 mA	0	0
R1	$I_{R1}R_1$	−(5 V)+	+(2.5 V)−	0	0
R2	$I_{R2}R_2$	+(5 V)−	−(2.5 V)+	0	0
6	$V_4 \pm V_{R2}$	−5 V	+7.5 V	0	−5 V
5	$V_3 \pm V_{R1}$	+15 V	−2.5 V	0	−5 V
8	$\tfrac{1}{2}V_6$	−2.5 V	+3.75 V	0	−2.5 V
7	$= V_8$	−2.5 V	+3.75 V	0	−2.5 V
R3	$\left\|\dfrac{V_5 - V_7}{R_3}\right\|$	→ 1.75 mA	← 0.625 mA	0	← 0.25 mA
R5	$= I_{R3}$	→ 1.75 mA	← 0.625 mA	0	← 0.25 mA
R5	$I_{R5}R_5$	+(17.5 V)−	−(6.25 V)+	0	−(2.5 V)+
9	$V_7 \pm V_{R5}$	−20 V	+10 V	0	0

if $R_1 = R_2$. For the specific resistance values we have shown for example purposes, the gain is

$$A_V = 1 + \frac{50 \text{ k}\Omega}{R_G}$$

IC instrumentation amplifiers are quite convenient to work with. Typically they have high input impedance and low output impedance, which is quite desirable. Gains up to 1000 are usually possible. Differential inputs are available; if not needed, merely ground one input and apply a single signal line to the other input terminal. If the output-signal polarity (that is, phase) is wrong, merely switch the input-terminal connections. When operated from a dual polarity (that is, ±V) supply voltage, no biasing is needed. As you may have already suspected, the IC instrumentation amplifier is a fine component to be acquainted with.

8.11   IC OPTICALLY COUPLED ISOLATORS

Optically coupled isolators (optical isolators or opto isolators, for short) are IC devices that make use of the optical properties of *PN* junctions. The optical properties of interest have already been considered in the photo diode (light-sensitive diode) and the light-emitting diode (LED). An optical isolator makes use of both these properties in a single IC package. In Figure 8.50(a) we see a transistor controlled by the optical properties of an LED and a photo diode. Take special note of the fact that there is no electric conducting path between the input and output terminals.

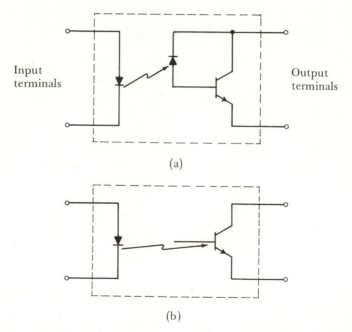

Figure 8.50   Optical-isolator circuits: (a) Photo-diode light receiver; (b) Photo-transistor light receiver.

In Figure 8.50(b), the light-receiving component is a photo transistor. A photo-transistor can be brought into a conducting state by an incident light beam in a manner similar to that of the photo diode we considered in Chapter 4. The similarity seems very real if we remember that the collector-base junction of a transistor is in fact a reverse-biased *PN* junction.

As implied by the name of the device, the main property of interest is isolation. Whereas the output signal needs to be a known function of the input signal, there should be no electric connection between input and output. This absence of an electric connection (that is, an electric conducing path) is what we call electrical isolation. The isolation property is important in several types of applications, which we will consider later. The ideal functional relationship between input and output would be for the two signals to be replicas of one another.

Now that we know something about an opto isolator, let us consider some applications in which its distinct properties are needed. In Figure 8.51 we see an opto isolator performing an interface

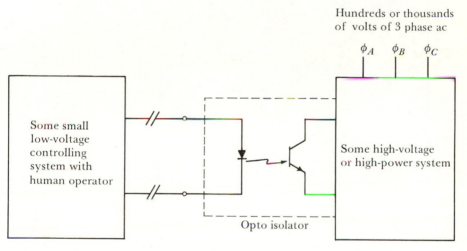

Figure 8.51   Optically coupled isolator application.

function between a high-voltage system and a low-voltage controlling system with a human operator. Without the isolating properties of the opto isolator, that is, if a conducting path existed between the systems, the human operator might be in imminent danger of a lethal electric shock.

Sometimes there is confusion about the voltage levels present in a situation such as that shown in Figure 8.51. For example, students sometimes assume that the transistor must have hundreds or thousands of volts between collector and emitter. To be sure that there is no misunderstanding on this topic, let us study the detailed examples shown in Figure 8.52. In (a) we have an opto isolator used to control an SCR in the high side of an ac line. Note that the maximum collector-emitter voltage across the photo transistor is 5 V even though the photo transistor controls a circuit with peak voltages

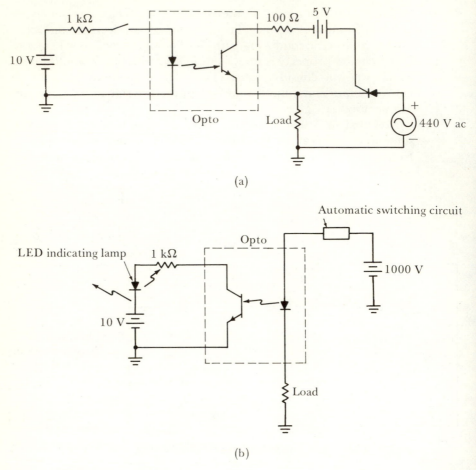

(a)

(b)

Figure 8.52 Detailed opto-isolator circuits: (a) isolated control circuit; (b) isolated indicator circuit.

greater than 600 V. More about ratings later. In Figure 8.52(b) we have the details of an opto isolator used to indicate the presence of current through a high-voltage load circuit. The LED will have only a normal forward-voltage drop across it and the photo transistor can have only a maximum of 10 V across it. The voltage level between photo transistor and LED (isolation voltage), however, will be about 1 kV.

Let us make some further observations about the circuits of Figure 8.52. Some students have probably already observed and concluded that if the SCR [part (a)] and the LED [part (b)] were moved and connected below the load resistor, next to the ground connection, no isolation would be needed. That is a keen and correct observation. There are cases, however, where for safety reasons one side of the load must be connected to ground. The examples shown are very realistic.

In addition to the circuits presented thus far, it is possible to buy optically isolated Darlington-connected transistors and optically isolated SCRs. The Darlington is preferred if a high-β transistor is needed. An opto SCR could be used in the circuit of Figure 8.52(a) to eliminate the need for the 5-V dc source and the 100-Ω resistor if a unit of sufficient voltage and current rating could be found. Opto SCRs are generally not available in high-voltage or high-current ratings.

A data sheet for an IC opto isolator is shown in Figure 8.53. Each parameter is well defined in terms of letter symbol and a word description. This data sheet should be self-explanatory. Two of the more important parameters are the current transfer ratio (CTR) and the isolation voltage. On this particular data sheet, the CTR is labeled as I_C/I_F, which is the exact definition. CTR is the ratio of transistor collector current to diode forward current.

The opto-isolator example applications we have considered thus far have been cases where the input and output signals need not be accurate replicas and where high isolation voltages were needed. There are still other cases, which we will consider briefly.

In Figure 8.54 we have a simple amplifier stage using an opto isolator. The Darlington-connected transistors are connected in a typical basic-amplifier configuration. Note that we cannot call this a common-emitter connection since there is no common connection between input and output. The input LED is biased by means of the battery and variable resistor. The fixed resistor is simply a precautionary device to protect the LED in case the pot should accidentally be set to zero resistance. The pot should be adjusted such that the output transistor has the desired Q point. The ac input signal is applied through the coupling capacitance C_1. The circuit as shown will have an ac voltage gain of about 10. It is anticipated that gain might vary substantially and the Q point move considerably if various 5082-4371 devices were substituted. This change would be because transistor and IC parameters change from unit to unit and this circuit has no self-adjusting (negative-feedback) properties. The output signal will be a reasonably good replica of the input-signal waveform.

Opto isolators can be used to advantage in medical-electronic circuits. In many of these cases the isolation voltage rating may not

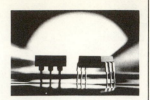

NCT200, NCT260 phototransistor opto-coupler

general description

The NCT200 and NCT260 are Gallium Arsenide diodes coupled with an NPN Silicon phototransistor in a six lead Epoxy dual-in-line package. These devices feature isolation voltage in excess of 2 kV. A GaAs light emitting diode radiates infrared light into a photo-sensitive transistor providing electrical isolation equivalent to a relay.

These devices are ideally suited where coupling is needed between two circuits but electrical isolation must be maintained. These devices find a wide range of application in data transmission as well as linear coupling.

applications

- Phase control
- Feedback control
- Telephone line receiver
- Line to digital logic isolation
- Solid state relays

features

- 2000V isolation
- High direct-current transfer ratio
- 0.5 pF coupling cap.
- Standard dual-in-line package

absolute maximum ratings

Storage Temperature	$-55°C$ to $+150°C$
Operating Temperature	$-55°C$ to $+100°C$
Total Power Dissipation at 25°C	250 mW
Derate Linearly	3.3 mW/°C
Lead Temperature (Soldering, 10 seconds)	260°C

output transistor ($T_A = 25°C$, $I_F = 0$)

Power Dissipation	200 mW
Derate Linearly from 25°C	2.6 mW/°C
V_{CEO}	30V
V_{CER} (1 MΩ)	70V
V_{ECO}	7V

input diode ($T_A = 25°C$)

Power Dissipation	200 mW
Derate Linearly from 25°C	2.6 mW/°C
Forward DC Current Continuous	60 mA
Forward DC Current Intermittent Duty*	150 mA
Reverse Voltage	3V
Peak Forward Current (1 pulse; 300 pps)	3A

*Dictated by maximum power dissipation.

physical dimensions

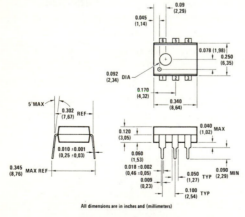

All dimensions are in inches and (millimeters)

Molded Dual-In-Line Package
Order Number NCT200 or NCT260

connection diagram

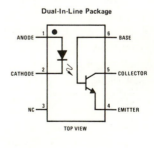

Figure 8.53 Opto-isolator data sheet. (Courtesy of National Semiconductor.)

electro-optical characteristics $(T_A = 25°C$, unless otherwise specified)

PARAMETER	CONDITIONS	MIN	TYP	MAX	UNITS
INPUT DIODE					
Forward Voltage (V_F)	$I_F = 60$ mA		1.2	1.5	V
Reverse Leakage Current (I_R)	$V_R = 3.0V$			10	µA
Capacitance (C)	$V = 0$, $f = 1$ MHz		150		pF
OUTPUT TRANSISTOR, $I_F = 0$					
DC Forward Current Gain (H_{FE}) (NCT200)	$V_{CE} = 10V$, $I_C = 1$ mA		500		
Collector-Emitter Current (I_{CER}) NCT200 NCT260	$V_{CE} = 10V$, $R = 1$ MΩ			50 100	nA nA
Collector-Emitter Sustaining Voltage (LV_{CEO})	$I_C = 1.0$ mA	30	60		V
Collector-Emitter Breakdown Voltage (BV_{CER})	$I_C = 100µA$, $R = 1$ MΩ	70	110		V
Collector-Base Breakdown Voltage (BV_{CBO})	$I_C = 10µA$	70	110		V
Emitter-Collector Breakdown Voltage (BV_{ECO})	$I_C = 100µA$	7	8.2		V
COUPLED CHARACTERISTICS					
DC Current Transfer Ratio (I_C/I_F) NCT200 NCT260	$I_F = 10$ mA, $V_{CE} = 10V$, $R_{BE} = 1$ MΩ	20 6	60		% %
Collector-Emitter Saturation Voltage $(V_{CE(SAT)})$ NCT200 NCT260	$I_F = 15.0$ mA, $I_C = 1.6$ mA $I_F = 50$ mA, $I_C = 6.4$ mA pulsed $I_F = 50$ mA, $I_C = 1.6$ mA		0.2	0.4 0.4 0.5	V V V
Isolation Voltage (V_{ISO})		2000			V
Isolation Resistance (R_{ISO})	$V_{ISO} = 500V$		10^{11}		Ω
Isolation Capacitance (C_{ISO})	$t = 1$ MHz		0.5		pF
Bandwidth (Note)	$I_F = 10$ mA, $V_{CC} = 5.0V$, $R_L = 100Ω$		150		kHz
Output On and Off Time t_{ON} t_{OFF}	$I_F = 10$ mA, $V_{CE} = 4.0V$, $R_L = 22Ω$		2.0 3.0		µs µs

Note: Bandwidth is specified as the point where the collector current transfer ratio is 1/2 that of the low frequency current transfer ratio (100 Hz).

switching time waveforms

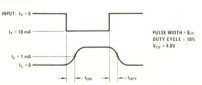

Manufactured under one or more of the following U.S. patents: 3083262, 3189758, 3231797, 3303356, 3317671, 3323071, 3381071, 3408542, 3421025, 3426423, 3440498, 3518750, 3519897, 3557431, 3560765, 3566218, 3571630, 3575609, 3579059, 3593069, 3597640, 3607469, 3617859, 3631312, 3633052, 3638131, 3648071, 3651565, 3693248.

National Semiconductor Corporation
2900 Semiconductor Drive, Santa Clara, California 95051, (408) 732-5000/TWX (910) 339-9240
National Semiconductor GmbH
808 Fuerstenfeldbruck, Industriestrasse 10, West Germany, Tele. (08141) 1371/Telex 27649
National Semiconductor (UK) Ltd.
Larkfield Industrial Estate, Greenock, Scotland, Tele. (0475) 33251/Telex 778-632

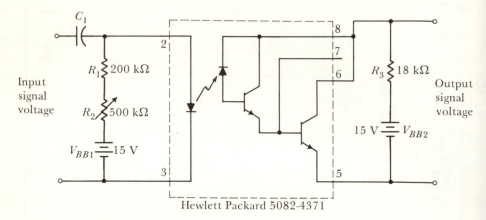

Figure 8.54 Simple amplifier circuit using an opto isolator.

need to be extremely high, but a high isolation resistance is mandatory. As the field of medicine gets more electrically oriented, electronic instruments perform more on-the-body or within-the-body functions. Some of the functions might be selecting signals (that is, separating common-mode and differential-mode signals) or amplifying signals. Whatever the exact application, it is important that the instrumentation be well isolated from the power-line voltages. The consequences of not maintaining isolation could be very serious.

8.12   ISOLATION AMPLIFIERS

Isolation amplifiers are amplifying circuits that maintain complete isolation between input signals and output signals. The isolation is commonly obtained by optical coupling or by transformer coupling.

Optically coupled isolation amplifiers are very closely related to the opto isolators we considered in Section 8.11. In fact the circuit shown in Figure 8.55 consists of two opto isolators and comes directly from a Hewlett-Packard application note (application note 939) for opto isolators.

Optically coupled isolation amplifiers are also available in single packages. An example would be the Burr-Brown 3650 family of isolation amplifiers which are shown pictorially in Figure 8.56(a), and whose block diagram representation is shown in (b). The dc/dc converter is not part of the isolation amplifier module. The dc output from the converter is isolated from the input. Whereas the

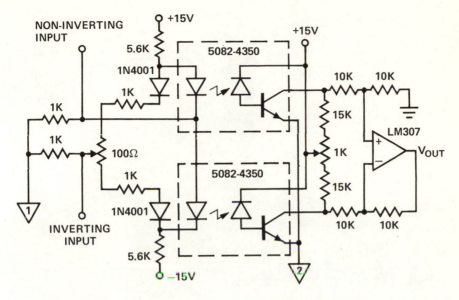

Figure 8.55 Isolation amplifier consisting of two opto isolators and associated components. (Courtesy of Hewlett Packard.)

modular form of isolation amplifier has the disadvantage of being larger than the IC, it may have specifications that are more desirable.

Transformer-coupled isolation amplifiers are available in modular packages. We would normally expect the transformer-coupled types to be physically larger than the optically coupled types. In Figure 8.57 we see a block diagram, with explanatory comments, of a certain family of transformer-coupled isolation amplifiers made by Analog Devices Corp. Note that only a single external supply-voltage source is needed. All other voltages needed internally are generated internally with proper isolation. The 125-kHz oscillator is used to modulate (chop) the signal and to demodulate the signal as well as generate an isolated dc voltage for the input stages.

A short discussion about the principles of chopper circuits seems appropriate at this time so that we can understand the block diagram of Figure 8.57. The blocks labeled modulator and demodulator are, in fact, chopper-type circuits.

Figure 8.58 shows a simple mechanical-switch version of the chopping modulator and demodulator needed for transformer-coupled isolation circuits. The chopper merely breaks up the signal into many segments such that each segment is reversed in polarity from the preceding segment. The resulting signal is totally ac, which can be transferred, with isolation, by a transformer. At the output of the

(a)

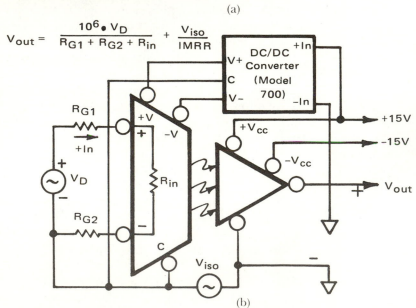

(b)

Figure 8.56   Module type of optically coupled isolation amplifier: (a) pictures of module packages; (b) block diagram of circuitry. (Courtesy of Burr-Brown Research Corporation.)

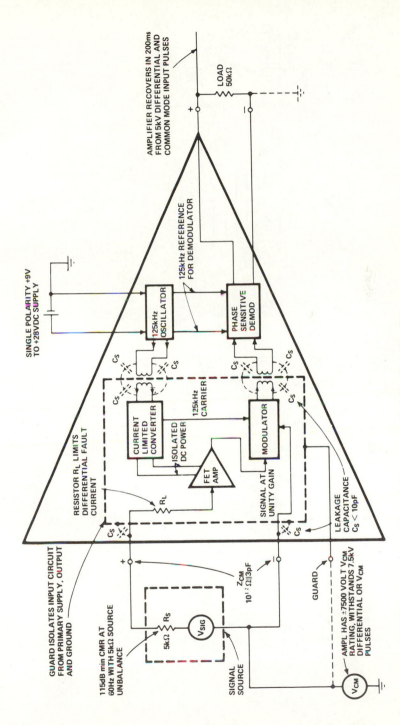

Figure 8.57 Block diagram of a transformer-coupled family of isolation amplifiers. (Courtesy of Analog Devices Corporation.)

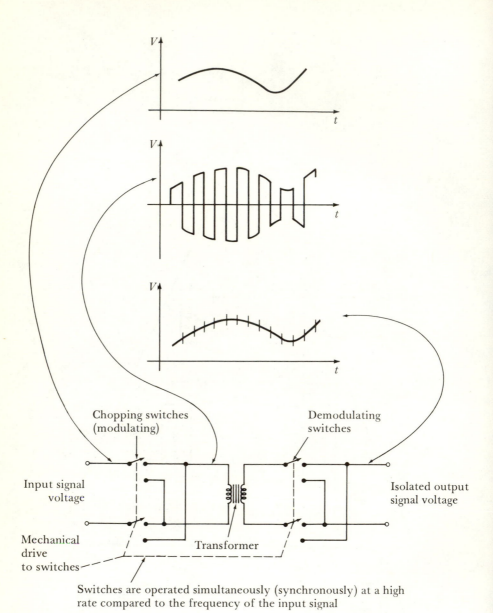

Figure 8.58   Circuit and associated waveforms of a simple chopper circuit.

transformer, another chopper is used to reassemble the signal in its original form. If both sets of switches operate simultaneously, every segment which is reversed by the modulating switches will be reversed again by the demodulating switches such that the input-signal waveshape is re-created at the output. During the process just described we have gained the isolation property but have introduced some distortion in the form of switching spikes (glitches) in the output-signal waveshape.

The discussion of chopper circuits is not meant to be complete. It is also not meant to imply that chopping must be done with mechanical make-and-break switches. In many cases the switching is done with transistors. The circuit was chosen because it is very simple. Hopefully it has been reasonably easy to follow the circuitry and line of reasoning we have used.

8.13 OSCILLATOR CIRCUITS

Oscillator circuits are electronic circuits that have an ac output signal without an ac input signal. Thus we see that oscillators are ac signal sources. We cover oscillators at this point because we prefer to explain and build oscillator circuits using previously covered ICs.

There are many types of oscillators. Some produce sinusoidal output-signal waveshapes and some produce nonsinusoidal wave-shapes. Nonsinusoidal waveshapes include square wave, rectangular wave, triangle wave, and sawtooth wave. Oscillator circuits can be built using BJTs, FETs, vacuum tubes, UJTs, tunnel diodes, differential amps, op amps, instrumentation amps, etc. There are many possible circuit connections for each of the types mentioned.

Our objective in this section, which may be difficult to achieve, is to instill in each student an intuitive understanding of the basic operating principles of oscillators without getting them involved in too many details. A thorough study of oscillators would be a major undertaking, which is not appropriate for this text.

As mentioned in our description of an oscillator, an external ac signal is not applied as an input in order to get an ac output signal. Since this is the case, it is not very difficult to believe that an oscillator supplies its own input signal. In fact, most oscillator circuits can be described as amplifier circuits where the signal input is self-generated. This self-generating or self-feeding effect is called *regeneration* or *positive feedback*.

Oscillators, of course, generate an ac output signal. However, in order to get a simple view of regeneration, let us consider a dc case.

Perhaps the simplest circuit we could connect to demonstrate regeneration is shown in Figure 8.59. Two CMOS amplifiers are connected in a cascade (signal-series) connection. Two stages are required because we need an overall noninverting amplifier. The intention in part (a) is to show that it is possible to set the pot so that we have equal input and output voltages. This condition will occur at near 7.5 V for the situation (+15 V supply) shown. Part (b) is intended to show that the overall circuit has a voltage gain of (typically) 400. The conditions depicted in parts (a) and (b) are important to our approach of explaining regeneration.

Based upon the conditions of Figure 8.59(a), where both input

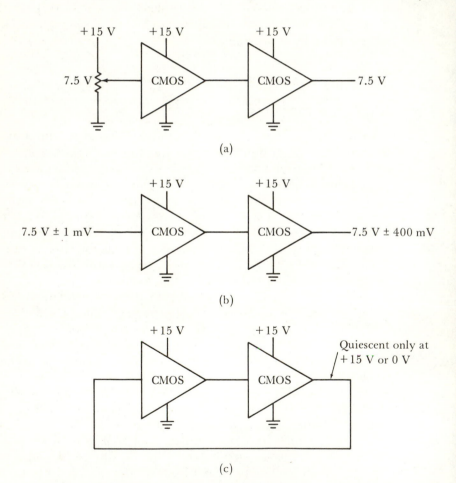

Figure 8.59 CMOS amplifiers used to illustrate regeneration (positive feedback): (a) quiescent operation; (b) typical input-output relationships; (c) regenerative quiescent states (there are two).

and output voltage are exactly equal, we might be prone to make a false assumption. It might seem possible for us to connect the output and input terminals together (they are at the same voltage), as in part (c), and have the quiescent input-output voltage remain at 7.5 V. However, we will show that it is not possible.

The conditions illustrated in Figure 8.59(b), namely, a voltage gain greater than 1 in a noninverting voltage-amplifying circuit, makes it impossible for the circuit of part (c) to have a quiescent input-output voltage of (or near) 7.5 V. The reasoning goes something like this. Assume that the circuit of part (c) is connected and does in fact have a quiescent input-output voltage of 7.5 V. Now if a disturbance of any kind caused the input voltage to increase slightly, say by 1 μV, the output would try to increase by 400 μV (remember the voltage gain is typically 400) and would in fact pull the input voltage up by 400 μV. As the input voltage increases, the output would tend to increase 400 times as much and pull the input up with it, and so forth. As you can see, this is a snowballing or self-feeding action we call regeneration. It will stop only when the amplifier operating point hits the end of its load line, in this case the supply voltage of 15 V. If our original disturbance had been in the decreasing direction, we would have had the same regenerative effect but with a final quiescent input-output voltage of (near) zero volts.

We stated earlier that in order to get regeneration, we would need a noninverting amplifier, that is, an amplifier with no polarity inversion. Since we have already shown that a noninverting amplifier will regenerate when the output signal is fed back to the amplifier input, let us now show that regeneration will not occur if an inverting amplifier is used. Figure 8.60 includes the circuits we will use in our discussion.

Figure 8.60(a) shows a simple CMOS amplifier stage. Note that the input and output terminals may be brought to the same dc voltage of 7.5 V. Also note that the amplifier has an inverting gain of (typically) 20. This is indicated by the expressions at the input and output terminals.

Figure 8.60(b) is a feedback connection. We want to consider the response of this circuit to a disturbance. Assume that the input voltage is disturbed by 1 mV in the positive direction. This would cause the output voltage to attempt to change by 20 mV in the negative direction. However, the output is connected to the input. As the output changes in the negative direction, it pulls the input voltage along with it. As soon as the output has gone 1 mV in the negative direction, the original disturbance has been canceled. Thus there is no cause for the output to change further. This connection tends to cancel any disturbances. It is called a *negative-feedback*

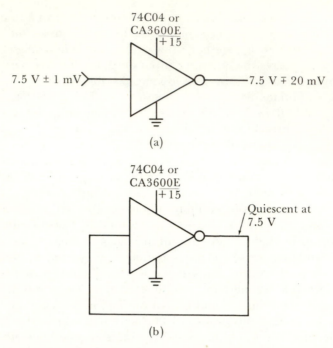

Figure 8.60   Circuits used to explain negative or degenerative feedback: (a) no feedback connection-inverting gain; (b) with feedback connection.

connection or a *degenerative-feedback connection*. This degenerating or canceling action is caused by the inverting action of the amplifier, that is, its polarity-inverting qualities.

We have now shown two conditions dealing with feedback around amplifiers. In the first case, Figure 8.59, we had regenerative feedback to the point where a runaway condition occurred. Any slight disturbance caused a regenerative runaway condition that halted only when the output voltage reached a limiting value. In other words, any disturbance was reinforced and grew in amplitude until a limiting value was reached. In the second case, Figure 8.60, we used an inverting amplifier that resulted in degenerative feedback. In this case, any disturbance was canceled and the original conditions were quickly restored.

In the case of a circuit with regenerative feedback, there are two distinct modes of operation. The mode of operation is a function of amplifier gain (in the simple case we are considering). If the amplifier gain is greater than 1, as in the circuit of Figure 8.59, then any disturbance grows in a runaway fashion until the circuit reaches a limiting value. If the amplifier gain is less than 1, any disturbance will be built up somewhat by the regenerative feedback but not to the point of a runaway situation. In Figure 8.61 we have two CMOS amplifiers connected in op-amp fashion. The net effect is a nonin-

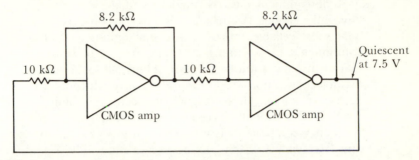

Figure 8.61   Regenerative amplifier connection.

verting amplifier with a voltage gain of less than 1. When the output is connected to the input, as in the figure, you will find that no runaway occurs and that the quiescent operating point is where the input-output voltage is near 7.5 V. This proves that regenerative (or positive) feedback will not cause a runaway condition when the amplifier gain is less than unity.

We have now concluded our nonmathematical and intuitive discussion of regeneration. Hopefully students will want to build the example circuits to prove to themselves the validity of our thoughts.

Now we need to consider regeneration in an ac case so that we can understand oscillators. Consider the block diagram of Figure 8.62. It

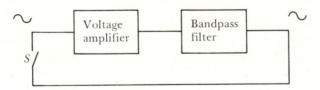

Figure 8.62   Block diagram showing simple concept of an oscillator.

consists of an amplifier and a bandpass filter. Assume that there is only one narrow band of frequencies where the signal voltage out of the bandpass filter is a replica of the input signal (in phase) and is larger than the input-signal voltage. Now suppose that some arbitrary disturbance causes a sine wave of the correct frequency to appear at the amplifier input terminal. A larger replica waveform will appear at the output of the filter circuit. If the switch is closed, the larger output signal will drag the input voltage to larger values that will increase output and, in turn, increase input, and so forth. The circuit will thus oscillate at the pass frequency of the bandpass filter.

Some comment must be made about the disturbance signals we have talked about. Disturbance signals can come from many sources. The most reliable source is thermal agitation. Unless all circuit components are operating at a temperature of absolute zero, thermal disturbances will be present. Thermal disturbances are caused by molecular action within the materials of which the circuit components are made. The disturbances occur randomly and are thus considered to contain all frequency components. The presence of all frequency components is important. It ensures that an oscillator of any frequency will tend to be self-starting. The signals resulting from thermal agitation are commonly called thermal noise signals. Thermal noise signals are small in terms of the expected output signal. However, they usually are enough to get some action started after which the oscillations build up in magnitude as a result of regeneration.

Now let us consider a simple, but buildable and operable, oscillator circuit. The oscillator circuit of Figure 8.63 will produce a

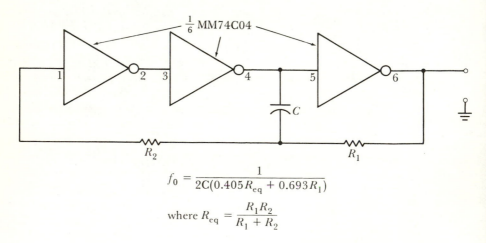

$$f_0 = \frac{1}{2C(0.405R_{eq} + 0.693R_1)}$$

$$\text{where } R_{eq} = \frac{R_1 R_2}{R_1 + R_2}$$

Figure 8.63 Oscillator made with three CMOS amplifiers (square-wave output). (Courtesy of National Semiconductor.)

square-wave output voltage whose peak-to-peak amplitude is equal to the supply voltage (between 3 and 15 V). The approximate frequency of oscillation can be calculated according to the equation given in the figure. This circuit can be made to operate from 0.1 Hz or less to about 1 MHz. This circuit is usually analyzed in terms of time constants instead of the amplifier and filter approach we used to explain regeneration. This is evident in the fact that the frequency of oscillation is given in terms of RC products. For applications where a square waveshape is sufficient, this circuit is small, economical, and easy to build.

The circuit shown in Figure 8.64 is similar to the block diagram

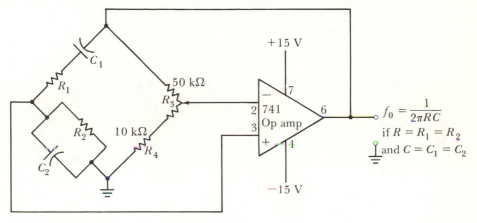

Figure 8.64   Oscillator made with op amp and bridge circuit (sinusoidal output waveshape—see text).

representation of an oscillator circuit shown in Figure 8.62. The amplifier in this case is an op amp and the filter circuit is an ac bridge. Note that the signal-output voltage of the amplifier drives the bridge and that the signal output of the bridge drives the differential inputs of the amplifier. The phase response of the bridge is all-important in this application as far as determining the frequency of oscillation. Only at one frequency, that is, at bridge-balance conditions, is the feedback signal (out of the bridge) of the correct phase to reinforce an original disturbance. Thus the circuit oscillates at a bridge-balancing frequency. The frequency of oscillation is given by the equation shown in the figure. The 2π factor in the f_0 equation indicates that the analysis was based on assumed steady-state sinusoidal conditions. This circuit will in fact produce a sinusoidal output voltage waveshape if the 50-kΩ pot is properly adjusted. The pot adjusts the magnitude of the regenerative-feedback signal. If too much regenerative feedback is used, it will cause the output-signal voltage to hit hard against the op-amp supply voltages (that is, the internal amps hit the end of their respective load lines) with resulting distortion of the output-signal waveform. If the pot is set so that significant distortion of the output waveform occurs, the frequency of oscillation will not be accurately indicated by the given f_0 equation.

Hopefully every reader now has an understanding of what an oscillator circuit is and also a good intuitive feel for what makes a circuit oscillate.

EXERCISES

QUESTIONS

Q8.1 What advantages and disadvantages should be noted for follower-type circuits?

Q8.2 Why are common-control-element amplifiers seldom used?

Q8.3 SCR characteristics seem well suited for use in ac circuits. Why?

Q8.4 What is the meaning, and significance of, the intrinsic stand-off ratio in UJT devices?

Q8.5 In what way is a PUT advantageous over the standard UJT?

Q8.6 What is a common application for UJTs?

Q8.7 What is an integrated circuit?

Q8.8 In what simplified manner have we treated ICs in this text?

Q8.9 What are some of the features that make CMOS amplifiers interesting and useful?

Q8.10 What is meant by voltage regulation?

Q8.11 How is it possible for local IC voltage regulators to reduce the cost of a main-system power supply?

Q8.12 Why would anyone ever want to amplify the difference between two signals?

Q8.13 Why is it appropriate to call certain amplifiers operational amplifiers?

Q8.14 How does an instrumentation amplifier differ from an operational amplifier?

Q8.15 Why are the words optical and isolator appropriate for the device given that name?

Q8.16 How does an isolation amplifier differ from an optical isolator?

Q8.17 What do we call an electronic circuit that has an ac output signal but no input ac signal?

Q8.18 What are some conditions necessary for a circuit to maintain a snowballing effect?

PROBLEMS

P8.1 Using ac-equivalent-circuit methods, find the input impedance for the circuit of Figure 8.65. The BJT parameters are $\beta = 75$, $r_b = 500\ \Omega$, and $V_{BB} = 20$ V.

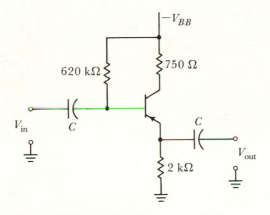

Figure 8.65   See Problems 8.1 and 8.2.

P8.2 Find the output impedance for the circuit of Figure 8.65 if $\beta = 100$ and $r_b = 1000\ \Omega$.

P8.3 Find the input impedance for the circuit shown in Figure 8.66 if $\beta = 150$ and $r_b = 1000\ \Omega$.

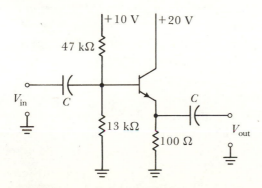

Figure 8.66   See Problems 8.3 and 8.4.

P8.4 Find the output impedance for the emitter follower shown in Figure 8.66 if $\beta = 50$ and $r_b = 500 \ \Omega$.

P8.5 For the Darlington circuit shown in Figure 8.67, calculate R_1 and R_2 values to give a quiescent collector current of 10 mA. Other specifications are $I_{R2} = 3I_b$, $\beta_1 = \beta_2 = 20$, silicon transistors.

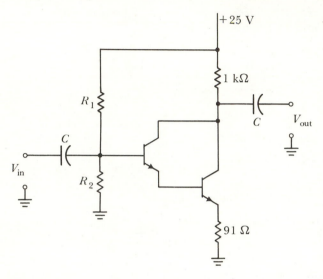

Figure 8.67   See Problems 8.5 and 8.6.

P8.6 Calculate the input impedance for the circuit of Figure 8.67 using ac equivalent circuit methods. The circuit parameters are as follows. $\beta_1 = \beta_2 = 50$, $R_2 = 330 \ \text{k}\Omega$, $R_1 = 750 \ \text{k}\Omega$, silicon transistors.

P8.7 Use ac-equivalent-circuit techniques to find the voltage gain of the Darlington amplifier circuit of Figure 8.68. The germanium PNP transistors are described as $\beta_1 = \beta_2 = 50$ and $r_{b1} = r_{b2} = 200 \ \Omega$.

P8.8 Find the input impedance for the circuit of Figure 8.68 where the transistors are as follows: $\beta_1 = \beta_2 = 50$, $r_{b1} = r_{b2} = 200 \ \Omega$.

P8.9 Repeat Problem 8.8 to find output impedance.

P8.10 Figure 8.69 shows a Darlington emitter-follower circuit. Use ac equivalent circuits to find the ac voltage gain. The transistors are identical and are specified as $\beta_1 = \beta_2 = 70$ and $r_{b1} = r_{b2} = 500 \ \Omega$.

P8.11 Find the ac input impedance for the circuit of Figure 8.69 if $\beta_1 = \beta_2 = 70$ and $r_{b1} = r_{b2} = 500 \ \Omega$.

P8.12 Repeat Problem 8.11 to find output impedance.

P8.13 The circuit of Figure 8.18(a) has the following numerical

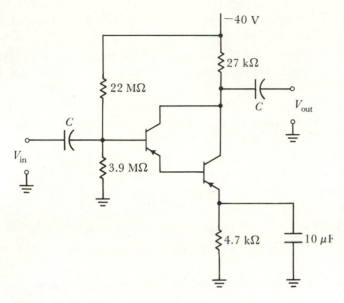

Figure 8.68   See Problems 8.7 and 8.8.

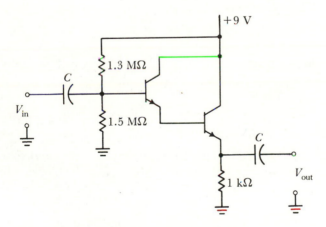

Figure 8.69   See Problems 8.10 and 8.11.

values: V_{in}(pp) = 500 V, R_L = 10 Ω. Find the current and voltage ratings needed for the SCR.

P8.14 Draw the circuit diagram and explain the operation of a UJT relaxation-oscillator circuit.

P8.15 Explain the internal operation of a CMOS amplifier circuit.

P8.16 Explain the one-resistor bias technique used for CMOS amplifiers.

P8.17 Describe a typical application for IC voltage regulators in a large system.

P8.18 A differential amplifier, operating in a single-ended mode, has the following specifications: $A_V(\text{DMSE}) = 200$, CMRR = 1000. Find the common-mode gain of the differential amplifier.

P8.19 A differential amplifier, operating in a differential-ended mode, has the following specifications: CMR = 0.01, CMRR = 2000. Find the gain $A_V(\text{DMDE})$ of the amplifier.

P8.20 Experimental testing on a particular differential amplifier yielded the following results: 1 V common-mode input gives 1-mV output (SE); 1-mV differential-mode input gives 1-V output (SE). Find $A_V(\text{CMSE})$, $A_V(\text{DMSE})$, and CMRR.

P8.21 Draw the block diagram of an op-amp integrating circuit.

P8.22 Derive the equation that describes a two-input summing op-amp circuit.

P8.23 Draw the op-amp circuitry necessary to form the equation $v_0 = v_1 + 5v_2 - 10v_3 - 2v_4$.

P8.24 Draw an op-amp differentiating circuit. Why are differentiators avoided if possible?

P8.25 Draw the op-amp circuitry necessary to form the equation $v_0 = 2v_1 + 2v_2 - 5v_3 + \int v_4 \, dt$

P8.26 For the instrumentation amplifier shown in Figure 8.49:

(a) Calculate the value of R_G required to produce a voltage gain of 20.

(b) Draw the circuit connection that will yield a noninverting amplifier with a gain of 20.

P8.27 If the R_G value shown in Figure 8.49 is changed to 10 kΩ, calculate all necessary voltages and currents, as in Table 8.1, and find output-voltage and differential-voltage gain. The input voltages are $V_1 = 0$ V, $V_2 = -1$ V.

P8.28 Show circuit connection and needed R_G value to obtain a single-ended inverting voltage gain of 33 with the instrumentation amplifier of Figure 8.49.

P8.29 Repeat Problem 8.31 with a circuit that will yield a differential-ended gain of 50.

P8.30 Describe some situations where optical isolators can be used to advantage.

P8.31 List some applications that need the characteristics of an isolation amplifier.

CHAPTER 9
RELATED TOPICS IN ELECTRONICS

9.1 THE DECIBEL AS A UNIT OF MEASURE

In some situations it is desirable to describe the magnitude of quantities in some nonlinear manner. This may be so when the quantity in question varies in magnitude over an extreme range. Another situation might be in some circumstance where a system responds in a nonlinear fashion to an excitation function.

The appropriate systems of the human body respond in a nonlinear fashion to excitations of light, sound, pain, taste, and so forth. The body is subjected to excitations in these areas that vary over tremendous ranges. The response is in most cases an approximation of a logarithmic function. The logarithmic function has the effect of making the response function vary over a much smaller range than the excitation function.

Consider the response of your body to excitations with which we are all familiar. Your body responds quite noticeably to the sound of a mosquito in the quiet of a still night. If our ears responded linearly to excitations of this order, 10 mosquitos would produce a deafening sound (response). On a dark night the human eye can easily distinguish the amount of light produced by the glowing hands of a watch or clock. If it were not for the logarithmic response of our sight system, a few watches or a few lightning bugs could supply all of our lighting needs. If our sight system were not logarithmic in response, we would not be able to venture outside in the daylight—not even with our eyelids closed.

In many cases in electronics we are interested in ratios. The gain of an amplifier, for example, is a ratio. Specifically, gain is the output quantity divided by the input quantity. Gain may be specified in the logarithmic manner we have just discussed.

Consider the case of power gain G. It may be calculated as

$$G = \frac{P_{out}}{P_{in}}$$

The power gain may also be calculated in bel units, which are indicated by the symbol B. This unit originated in the telephone industry and is named in honor of Alexander Graham Bell. By definition,

$$G = \log_{10} \frac{P_{out}}{P_{in}} \quad B$$

In many electronic circuits, such as amplifiers, we are interested in calculating the gain. Since gain is a ratio of output to input, the decibel is a convenient unit for expressing gain. Assume that we have an amplifier whose signal input power is 10 mW and whose signal output power is 150 mW. The power gain is

$$G = \frac{150 \text{ mW}}{10 \text{ mW}} = 15$$

$$= 10 \log_{10} \frac{P_{out}}{P_{in}} \text{ dB} = 10 \log_{10} \frac{150 \text{ mW}}{10 \text{ mW}} \quad \text{dB}$$

$$= 10 \log_{10} 15 \text{ dB} = 10(1.176) \text{ dB} = 11.76 \text{ dB}$$

The bel unit (and the decibel) is not restricted to the case of gain applications. It applies in any case where a ratio of powers is involved. In the more general case we have,

$$G = \log_{10} \frac{P_2}{P_1} B = 10 \log_{10} \frac{P_2}{P_1} \quad \text{dB}$$

P_2 and P_1 represent any power quantities whose ratio is of interest.

In some cases the numerator power magnitude may be smaller than the denominator term. When evaluated numerically, the resulting quantity will be negative. As an example,

$$G = 10 \log_{10} \frac{P_2}{P_1} \text{ dB} = 10 \log_{10} \frac{5 \text{ W}}{10 \text{ W}} \quad \text{dB}$$

$$= 10 \log_{10} \left(\frac{1}{2}\right) \quad \text{dB} = -3.01 \text{ dB}$$

This situation may be referred to as attenuation (α) rather than a gain. Mathematically,

$$\alpha = 10 \log_{10} \frac{2}{1} \quad dB = 3.01 \ dB$$

In many instances we are concerned with the voltage gain or current gain of various circuits. Using known relationships, we can convert from units of power to units of either voltage or current. The resistance across which the input power is developed and the resistance across which the output power is developed will appear in the expression. The conversion, in terms of voltage, is as follows:

$$G = 10 \log_{10} \frac{P_{out}}{P_{in}} \quad dB$$

$$= 10 \log_{10} \frac{V_{out}^2 / R_{out}}{V_{in}^2 / R_{in}} \quad dB$$

$$= 10 \log_{10} \frac{V_{out}^2}{V_{in}^2} \frac{R_{in}}{R_{out}} \quad dB$$

$$= 10 \log_{10} \frac{V_{out}^2}{V_{in}^2} + 10 \log_{10} \frac{R_{in}}{R_{out}} \quad dB$$

$$= 20 \log_{10} \frac{V_{out}}{V_{in}} + 10 \log_{10} \frac{R_{in}}{R_{out}} \quad dB$$

The conversion, in terms of current, is accomplished similarly.

$$G = 20 \log_{10} \frac{I_{out}}{I_{in}} + 10 \log_{10} \frac{R_{out}}{R_{in}} \quad dB$$

Note the opposite relation of the resistances in the second term of the current calculation as compared with the voltage calculation.

When dealing with measurements in units of decibels, we find that common usage in some cases differs noticeably from the definition. It is very common, though not true in the strict theoretic sense, to neglect the term dealing with input and output resistances when speaking of voltage and current gain. Thus we commonly say:

$$A_V = 20 \log_{10} \frac{V_{out}}{V_{in}} \quad dB \quad \text{and} \quad A_I = 20 \log_{10} \frac{I_{out}}{I_{in}} \quad dB$$

What we are really assuming is that the input and output resistances are equal.

EXAMPLE 9.1   A certain amplifier circuit has the following resistances and signal levels.

$$R_i = 5 \text{ k}\Omega$$

$$R_L = 1 \text{ k}\Omega$$

$$V_{\text{in}} = 10 \text{ mV}$$

$$V_{\text{out}} = 100 \text{ mV}$$

Calculate the power gain in terms of signal voltages and in terms of signal currents. Calculate voltage gain and current gain. All gains are to be in decibels.

SOLUTION   In terms of voltages,

$$G = 20 \log_{10} \frac{V_{\text{out}}}{V_{\text{in}}} + 10 \log_{10} \frac{R_{\text{in}}}{R_{\text{out}}} \quad \text{dB}$$

$$= 20 \log_{10} \frac{100}{10} + 10 \log_{10} \frac{5 \text{ k}\Omega}{1 \text{ k}\Omega} \quad \text{dB}$$

$$= 20 \, (1) + 10(0.7) \text{ dB}$$

$$= (20+7) \text{ dB}$$

$$= 27 \text{ dB}$$

In terms of currents,

$$G = 20 \log_{10} \frac{I_{\text{out}}}{I_{\text{in}}} + 10 \log_{10} \frac{R_{\text{out}}}{R_{\text{in}}} \quad \text{dB}$$

We must calculate I_{OUT} and I_{IN} from voltage and resistance data:

$$I_{\text{out}} = \frac{V_{\text{out}}}{R_L} = \frac{100 \text{ mV}}{1 \text{ k}\Omega} = 100 \, \mu\text{A}$$

$$I_{\text{in}} = \frac{V_{\text{in}}}{R_{\text{in}}} = \frac{10 \text{ mV}}{5 \text{ k}\Omega} = 2 \, \mu\text{A}$$

$$G = \left(20 \log_{10} \frac{100 \, \mu\text{A}}{2 \, \mu\text{A}} + 10 \log_{10} \frac{1 \text{ k}\Omega}{5 \text{ k}\Omega} \right) \text{dB}$$

$$= [20 \log_{10} 50 + 10 \log_{10} (0.2)] \text{ dB}$$

$$= [20(1.7) + 10(-0.7)] \text{ dB}$$

$$= (34 - 7) \text{ dB}$$

$$= 27 \text{ dB}$$

Voltage gain:

$$A_V = 20 \log_{10} \frac{V_{\text{out}}}{V_{\text{in}}} = 20 \log_{10} \frac{100 \text{ mV}}{10 \text{ mV}} \text{ dB}$$

Table **9.1**   Relationships between gain expressed in linear unitless quantities and as expressed in decibel units.

Linear power gain	Power gain, dB	Linear V or I gain	V or I gain, dB
1	0	1	0
2	3	2	6
$\frac{1}{2}$	-3	$\frac{1}{2}$	-6
4	6	4	12
$\frac{1}{4}$	-6	$\frac{1}{4}$	-12
8	9	8	18
$\frac{1}{8}$	-9	$\frac{1}{8}$	-18
16	12	16	24
$\frac{1}{16}$	-12	$\frac{1}{16}$	-24
10	10	10	20
$\frac{1}{10}$	-10	$\frac{1}{10}$	-20
20	13	20	26
$\frac{1}{20}$	-13	$\frac{1}{20}$	-26
40	16	40	32
$\frac{1}{40}$	-16	$\frac{1}{40}$	-32

$$= 20 \log_{10} (10) = 20(1) \text{ dB}$$

$$= 20 \text{ dB}$$

Current gain:

$$A_I = 20 \log_{10} \frac{I_\text{out}}{I_\text{in}} = 20 \log_{10} \frac{100 \ \mu A}{20 \ \mu A} \ \text{dB}$$

$$= [20 \log_{10} (5)] = 20(0.7) \text{ dB}$$

$$= 14 \text{ dB}$$

Some relationships between linear gain expressions and gain as expressed in decibels are shown in Table 9.1.

9.2 *R-C* COUPLING

By the use of resistive and capacitive elements, we are able to differentiate between, and therefore treat differently, ac signals and dc signals. A common application of the *R-C* coupling network is where ac signals are allowed to pass from one point to another in a circuit while dc signals are prevented from passing between the same two points. In Figure 9.1 we see an *R-C* coupling network with inputs and outputs shown. Note that the transmission between input and output (N is neutral or ground, V_{AN} is input, V_{BN} is output) is very much different for the dc case as compared with the ac case.

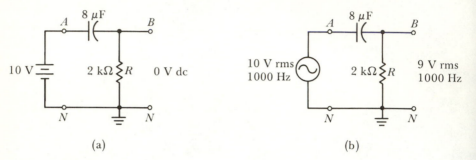

(a)   (b)

Figure 9.1   *R-C* coupling circuits: (a) response to dc input; (b) response to ac input.

The circuit of Figure 9.1 definitely does differentiate between dc and ac signals. When V_{AN} is a dc level, V_{BN} is zero. When V_{an} is a 1000-Hz signal voltage, about 90% of the V_{an} voltage magnitude level will appear as V_{bn}. The portion of the V_{an} signal voltage level that appears as V_{bn} is a function of the frequency of the V_{an} signal and the capacity (hence the X_c) of the coupling capacitor.

In order to choose a component value for C_C, we will need to know the frequency of the ac signal to be applied and what percentage of the ac voltage V_{an} we want to have present as V_{bn}. Mathematically we have

$$V_{bn} = V_{an} \frac{R}{R - jX_C}$$

or

$$\left| \frac{V_{bn}}{V_{an}} \right| = \left| \frac{R}{R - jX_C} \right|$$

Assuming we know the V_{bn}/V_{an} ratio and value of R, we can solve for the needed value of X_c. In order to find a coupling capacitor that will yield the proper X_c value, we must know the signal frequency. In cases where a range of frequencies may be present, use the lowest frequency in the X_c calculation. By doing so, the circuit will pass all higher frequencies with even less attentuation than the calculated V_{bn}/V_{an} ratio.

The dc blocking ability of an *R-C* coupling network is quite important. In Figure 9.2 we see a typical transistor amplifier stage

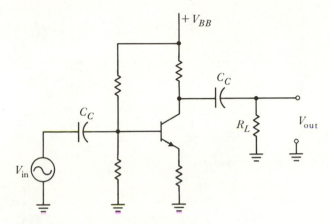

Figure 9.2 *R-C coupled amplifier.*

with *R-C* coupling at the input and output. The input-coupling capacitor keeps the signal source from disturbing the transistor dc bias levels. The output-coupling capacitor removes a dc level from the amplified signal before the signal is developed across the load resistance R_L.

9.3 CASCADE CONNECTION

In many cases we find that one stage of amplification is not sufficient. In such cases we usually use a cascade connection where the output signal of one stage becomes the input signal for another stage.

In Figure 9.3 we have a cascade connection of three amplifier stages. Interstage coupling is by *R-C* coupling networks. Note that dc blocking is all important in a connection such as this.

Consider the cascade amplifier connection shown in Figure 9.4. This could be a block diagram representation of the circuit of Figure 9.3. Let us assume that the voltage gain of each stage is 5 and the input signal level is 1 mV. As shown in the figure, the output of the first stage is 5 mV, the output of the second stage is 25 mV, and the output of the third stage is 125 mV. The voltage gain of each stage is seen to be 5, since the output is in each case five times larger than the input. The overall gain of the three-stage cascade connection is

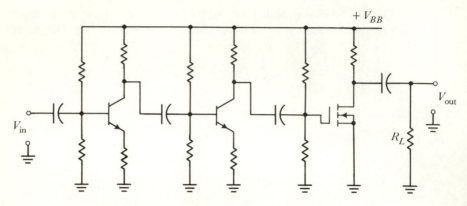

Figure 9.3   Cascade connection of amplifier stages.

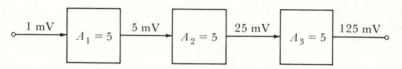

Figure 9.4   Block diagram of cascade connection of amplifier stages.

$$A_V = \frac{V_{\text{out}}}{V_{\text{in}}} = \frac{125 \text{ mV}}{1 \text{ mV}} = 125 \text{ mV}$$

In general, for any cascade connection of N stages, we find the overall gain to be

$$A_V = A_1 \cdot A_2 \cdot A_3 \cdot A_4 \cdots A_N$$

It is quite easy to obtain a large voltage gain by cascading the necessary number of stages.

The individual stage gains and the overall gain may be expressed in units of decibels. In this case the overall gain is written as

$$A_V \text{ (dB)} = A_1 + A_2 + A_3 + A_4 + + + A_N$$

Since we have just considered R-C coupling and cascading of amplifier stages, it seems appropriate to mention a theorem from circuit theory. The theorem is the maximum-power-transfer theorem. Students sometimes try to misapply the theorem to a situation where they want to transfer signals from stage to stage in a cascade connection of transistors or vacuum tubes. The theorem is valid, but we are not interested (normally) in transferring maximum power from stage to stage. Since FETs and vacuum tubes are voltage-controlled devices, we want to transfer maximum signal voltage to the input of a following stage. BJTs are current-controlled devices, so we want to transfer maximum signal current to the input of a BJT stage of amplification. However, a maximum signal voltage to the input of a BJT will result in a maximum signal current, so we generally say that our coupling circuits should transfer maximum voltage between stages. The desired maximum voltage will be transferred between stages when stage output impedances are minimized and stage input impedances are maximized.

9.4   AMPLIFIER FREQUENCY RESPONSE

The basic-amplifier-circuit connection we have considered is quite capable of amplifying ac signals. This is true whether we use a field-effect transistor, a bipolar-junction transistor, or a vacuum tube as the active device. The response of any of these amplifiers will necessarily be some function of frequency. Most amplifiers will have some particular voltage gain over a certain range of frequencies called the *midband frequency range*. The gain normally gets lower as the applied frequency gets above or below the midband frequency range.

Let us consider again some of the basic-amplifier-circuit connections. In particular we want to consider any reasons why the voltage gain should be a function of frequency. Since the gain decrease at high frequencies is a result of a different phenomenon than the gain decrease at low frequencies, we will consider the two cases independently.

In Figure 9.5 we have an FET basic-amplifier-circuit connection.

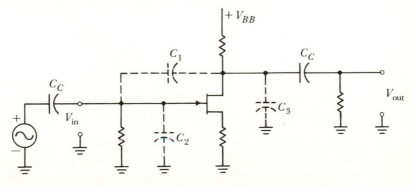

Figure   9.5   FET   basic-amplifier   circuit   showing   important   stray capacitances.

Note all the extra capacitances that we have never noted before. The capacitances shown in dashed lines are called *stray capacitances*. They are usually undesirable but cannot be eliminated entirely. Capacitances C_2 and C_3 are stray wiring capacitances. Since the connecting wires are conductors, the chassis is a conductor, and the two are separated by insulating material, there is some (hopefully small) capacitance between any wire and chassis. Capacitance C_1 is an interelectrode capacitance of the active device. Capacitance C_1 might be, in general, a drain-to-gate capacitance, a plate-to-grid capacitance, or a collector-to-base capacitance. In active devices meant for high-frequency operation, the interelectrode capacitance is minimized but can never be eliminated. Stray wiring capacitances can be minimized, but not eliminated, by careful and thoughtful construction practices.

Let us consider the consequences of the stray capacitances shown in Figure 9.5. We know from basic circuit theory that the X_C of any capacitance is a function of frequency. The greater the frequency, the smaller the X_C value. Although the stray capacitances are small, perhaps just a few picofarads or a fraction of a picofarad, there will be frequencies at which the reactance X_C of these capacitances will become small. The small value of X_C will significantly affect the operation of the circuit. Let us think about a frequency high enough so that C_1, C_2, and C_3 are essentially short circuits ($X_C = 0$). Under these assumptions, we find that the gate is shorted to ground, the drain is shorted to ground, and the drain is shorted to the gate. The circuit will have no gain at such a high frequency.

We have just shown that stray capacitances and interelectrode capacitances can affect the operation of amplifier circuits at sufficiently high frequencies. We will find that the stray and interelectrode capacitances have no effect at lower and midfrequencies. As frequencies are gradually increased above midband, we will find that the gain gradually decreases. At infinite frequency, the gain would go to actual zero, as discussed in the previous paragraph.

The low-frequency gain of an R-C-coupled amplifier is determined by the coupling capacitors. In Figure 9.6 we have an R-C-coupled bipolar-transistor basic-amplifier circuit. The reactance X_c of the coupling capacitors is of course a function of frequency. The coupling capacitors and their respective load resistances form a voltage-divider network. The equivalent load resistance for the input coupling capacitor consists of the transistor bias resistors and the input resistance of the transistor. The load resistor for the output coupling capacitor is of course R_L At low frequencies, where X_c is high, we get a lesser portion of voltage across the load resistances of the voltage-divider networks.

Now let us compare the gain of the amplifier of Figure 9.6 at low

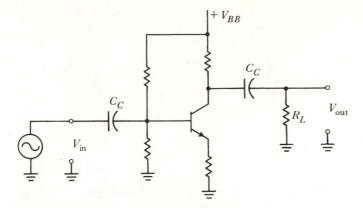

Figure 9.6   *R-C*-coupled transistor amplifier.

and midfrequencies. The coupling capacitors are essentially short circuits at midband. If the voltage gain of the amplifier from base to collector is 5, then the voltage gain from V_{in} terminals to V_{out} terminals will be essentially 5. As an example of low frequencies, we choose a frequency such that the voltage across the R part of the voltage divider is one-half the input voltage. We assume an ac input voltage of 50 mV. Under these assumptions, we find that we have a 25-mV signal at the base of the transistor. Since the voltage gain of the amplifier from base to collector is 5, we have 125 mV of signal voltage at the collector of the transistor. Assuming that the load resistance and output-coupling capacitor give us the same voltage division as the input *R-C* voltage divider, we will have 62.5 mV of signal at the V_{out} terminals. The gain of the circuit from V_{in} terminals to V_{out} terminals, at the low frequency chosen, is

$$A_v = \frac{V_{out}}{V_{in}} = \frac{62.5 \text{ mV}}{50 \text{ mV}} = 1.25$$

The voltage gain is definitely shown to decrease at low frequencies.

The frequencies at which the gain falls to 0.707 (that is, $1/\sqrt{2}$) of the midfrequency gain are called the half-power points or half-power frequencies. A decrease to 0.707 of a voltage or current level results in a decrease to 0.5 of the power level ($0.707^2 = 0.5$). Since half power represents a 3-dB change, the half-power points are also called the 3-dB points.

Many amplifier circuits have a frequency-response curve similar to the one shown in Figure 9.7. The lower and upper 3-dB frequencies are labeled f_1 and f_2, respectively. This particular graph is normalized so that midband gain is unity. A knowledgable circuit designer can design an amplifier so that the frequencies corresponding to f_1

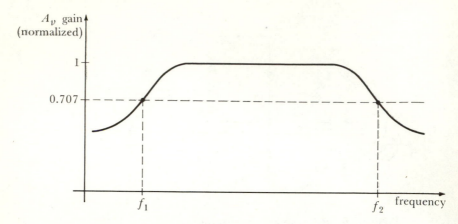

Figure 9.7   A typical frequency-response curve.

and f_2 will be as chosen. Whereas f_1 can go to zero frequency in the case of a dc-coupled amplifier (that is, an amplifier without coupling capacitors), we know of no way to extend f_2 to infinity.

The frequency-response curve of Figure 9.7 is valid for any of the R-C-coupled amplifiers we have discussed in this book. Sometimes it might be of interest to know the gain at frequencies above f_2 or the gain at frequencies less than f_1. These particular relationships have been worked out as follows:

$$|A_{\text{low}}| = A_{\text{mid}} \left|\frac{1}{1 - j\,(f_1/f)}\right|$$

$$|A_{\text{high}}| = A_{\text{mid}} \left|\frac{1}{1 + j\,(f/f)_2}\right|$$

The symbol A_{low} refers to the voltage gain at any frequency f that is lower than f_1. The symbol A_{high} refers to the voltage gain at any frequency f that is higher than f_2.

9.5 BANDWIDTH OF CASCADED AMPLIFIERS

We have already considered the overall gain of cascaded amplifier stages. We found that for N cascaded stages, the gain was $A = A_1 \cdot A_2 \cdot \cdot \cdot A_N$. The overall gain is seen to be the product of the individual stage gains.

Let us consider the overall bandwidth of cascaded stages. The result of this consideration may be surprising.

In Figure 9.8 we have the frequency-response curve of a particular amplifier circuit. We want to connect two such amplifier circuits in a cascade connection. Since the gain of each individual amplifier

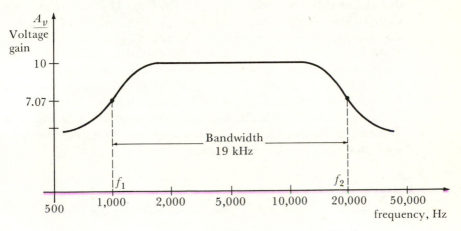

Figure 9.8   Frequency-response curve (gain versus frequency).

stage is a function of frequency, the overall gain of the cascaded connection will be a function of frequency. Calculating the overall gain (A_{OA}) at midband and at f_1 and f_2, we obtain the following data:

$$= 10 \times 10 = 100$$

$$A_{OA} \,(1000 \text{ Hz}) = 7.07 \times 7.07 = 50$$

$$A_{OA} \,(20 \text{ kHz}) = 7.07 \times 7.07 = 50$$

Immediately we see that the overall gain is down to less than 0.707 of peak at the former half-power frequencies (1000 and 20,000 Hz). Thus the bandwidth of the cascaded connection is less than the bandwidth of the individual stages.

Let us now make necessary calculations to determine the bandwidth of the cascaded stages. We have already determined that the midband gain of the cascaded connection is 100. We can calculate the gain of the cascaded connection at the 3-dB points to be $A_{OA} = 0.707 \times 100 = 70.7$. Since the overall gain A_{OA} is the product of individual stage gain, we have $A_{OA} = 70.7 = A_1 A_2$. But $A_1 = A_2$, so

$$A_{OA} = 70.7 = A_1^2$$

$$A_1 = A_2 = 8.42$$

Substituting the value of 8.41 into our equation for A_{low} and A_{high}, we can find the frequencies at which this gain will occur. We use the A_{low} and A_{high} equations here even though we previously specified that these equations are usually used for frequencies below f_1 or above f_2. From the A_{low} equation we obtain $F_1 = 1.550$ kHz. From the A_{high} equation we obtain $f_2 = 12.900$ kHz. The two frequencies we have just calculated are the f_1 and f_2 frequencies of the cascaded connection. At these frequencies the overall gain is 0.707 of the midband gain. The bandwidth of the cascaded connection is the difference between the new 3-dB points. Thus B = bandwidth = $f_2 - f_1$ = (12.9 − 1.55) kHz = 11.35 kHz. This represents a considerable reduction in bandwidth. The addition of more stages in a cascade connection will further reduce the overall bandwidth. The decrease in bandwidth for a fourth stage, for example, would be much less than for the second stage.

EXERCISES

QUESTIONS

Q9.1 Why is it sometimes advantageous to use a logarithmic system of measure such as the decibel system?

Q9.2 What is meant by the term cascade connection when referring to amplifier circuits?

Q9.3 What are the advantages and disadvantages of using R-C coupling between cascaded stages?

Q9.4 What causes the voltage gain of an amplifier circuit to fall off at sufficiently high frequencies?

Q9.5 What causes the voltage gain of R-C-coupled amplifier stages to decrease at low frequencies?

Q9.6 Why is the bandwidth of cascaded amplifiers less than that of the individual amplifier stages?

PROBLEMS

P9.1 A certain amplifier circuit has the following characteristics.

$R_{\text{in}} = 50$ kΩ $\qquad R_{\text{out}} = 2$ kΩ

$P_{\text{in}} = 1.25$ mW $\qquad P_{\text{out}} = 25$ mW

Find the power gain A_p in linear units and in decibels.

P9.2 Find the true power gain in decibels for the system described as follows.

$R_{in} = 1 \text{ M}\Omega$ $\qquad R_{out} = 50 \text{ k}\Omega$
$V_{in} = 10 \text{ mV}$ $\qquad V_{out} = 3.2 \text{ V}$

P9.3 For the system of Problem 9.2, find the nominal (that is, neglecting impedance differences) power gain in decibels.

P9.4 An amplifier has the following input and output quantities:

$R_{in} = 100 \text{ k}\Omega$ $\qquad R_{out} = 1 \text{ k}\Omega$ $\qquad R_L = 1 \text{ k}\Omega$
$V_{in} = 200 \text{ mV}$ $\qquad V_{out} = 7 \text{ V}$

(a) Find the voltage gain in linear units.

(b) Find the voltage gain in decibel units.

P9.5 Find the current gain for the amplifier of Problem 9.4.

(a) In linear units $\qquad$ (b) In decibel units

Note: currents can be determined from known voltage and impedance levels.

P9.6 A variable gain system has an input signal of 10 mW. Find the power gain in decibels when the output signal is

(a) 25 mW $\qquad$ (b) 50 mW

(c) 100 mW $\qquad$ (d) 200 mW

Note how gain in decibels varies as power gain doubles.

P9.7 Design a basic-amplifier circuit using either a BJT or an FET as the active device. Now draw the schematic diagram of a two-stage R-C-coupled cascade amplifier using your amplifier as the individual-stage amplifier circuit. Don't forget the coupling capacitors at input and output.

P9.8 Use ac-equivalent-circuit techniques to find the voltage gain of the single amplifier stage you designed in Problem 9.7.

(a) In linear units $\qquad$ (b) In nominal decibel units

P9.9 Find the voltage gain of the cascaded amplifier you designed in Problem 9.7. Use ac equivalent circuits.

(a) In linear units $\qquad$ (b) In nominal decibel units

P9.10 A certain amplifier has a midband gain of 100. Its half-power frequencies are f_1, 20 Hz, f_2, 20 kHz. Find the gain at the following frequencies.

(a) 10 Hz $\qquad$ (b) 30 kHz

ANSWERS TO SELECTED EXERCISES

P0.1 $I_{R1} = \overrightarrow{1 \text{ A}}$

$I_{R2} = 0.5 \text{ A} \downarrow$

$I_{R3} = 0.5 \text{ A} \downarrow$

P0.3 $V_{AB} = -5 \text{ V}$

$V_{ab} = \dfrac{3}{\sqrt{2}} \text{ V} = 2.121 \text{ V}$

$v_{AB} = (-5 + 3 \sin \omega t) \text{ V}$

$v_{ab} = (3 \sin \omega t) \text{ V}$

P0.5 $V_{FB} = V_{AB} + V_{GA} + V_{FG}$

P2.1

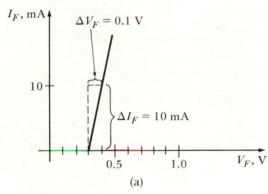

(a)

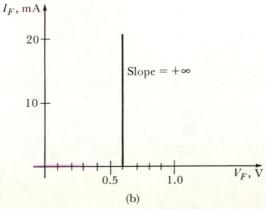

(b)

P2.3 Assuming ideal diode,

$$I_{\text{peak}} \text{ (min)} = 50 \text{ mA}$$
$$I_{\text{av}} \text{ (min)} = 15.92 \text{ mA}$$
$$PIV \text{ (min)} = 100 \text{ V}$$

P2.5 $V_F \approx 0.3$ V
$r \approx 14 \ \Omega$

P2.7 (a) Slightly altered half-wave-rectified sine wave whose peak value is 47.33 mA

(b) Slightly altered half-wave-rectified sine wave whose peak value is 94.67 volts

(c) Small errors will result

P2.9 No current or output voltage

P2.11

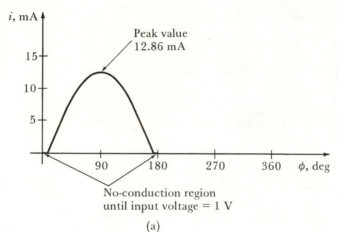

(a)

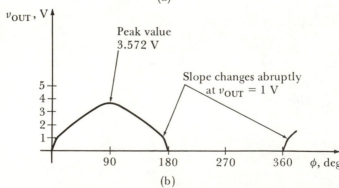

(b)

P2.13 (a) A square wave whose peak values are 0.0 mA and 18.48 mA counterclockwise

(b) A square wave whose peak values are 0.0 V and −18.48 V

P2.15 (a) A square wave whose peak values are 5.900 mA and 1.967 mA
(b) A square wave whose peak values are 29.5 V and 9.5 V

P2.17

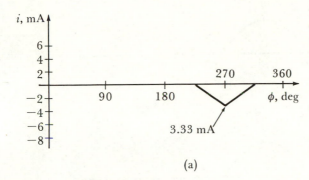

(a)

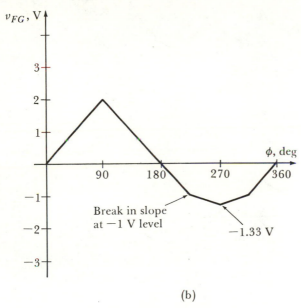

(b)

P2.19 (continues on p. 390)

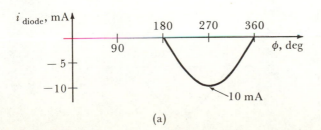

(a)

P2.19 (continued)

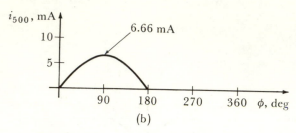

(b)

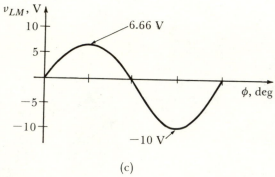

(c)

P2.21

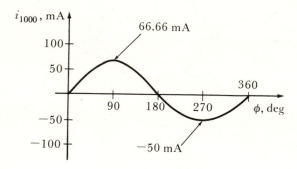

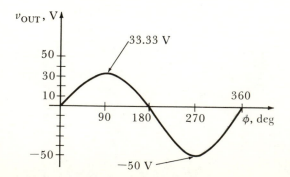

P3.1 See text
P3.3

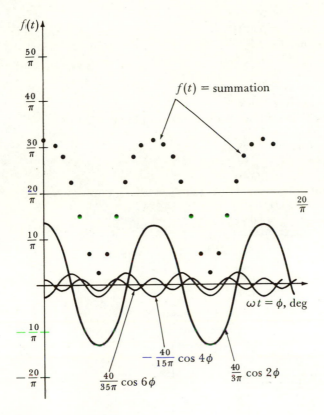

P3.5 (a) 13.23 V_{pp}
     (b) 158.85 V
P3.7 (a) 2.1 V_{pp}
     (b) 164.5 V
P3.9 32.49 kΩ
P3.11 59.29 μF
P3.13 395.3 μF
P3.15 (a) 4088 μF
     (b) 8.81 V
P3.17 (a) 108.7 μF
     (b) 163 V
P3.19 (b) 65.84 mV
     (c) 7.958 V
P3.21 (b) 570 mV
P4.1 R_1 (max) = 764.7 Ω
     P_{R1} (max) = 1888 mW
     V_z = 12 V
     P_z (max) = 452.3 W

P4.3 R_1 (max) = 84.34 Ω
P_{R1} (max) = 581 mW
V_Z = 8 V
P_Z = 584 mW

P4.5 R_1 (max) = 500 Ω
P_{R1} (max) = 1.8 W
V_Z = 20 V
P_Z = 1.2 W

P4.7 V_{IN} (max) = 160 V
V_{IN} (min) = 121 V

P4.9 The following values based upon I_Z (min) = 2 mA:
R_1 (max) = 2.083 kΩ
P_{R1} (max) = 2.028 W
V_Z = 50 V
P_Z (max) = 1.06 W
Most designs should use standard values, thus
R_1 = 2.0 kΩ, P_{R1} = 5 W, P_Z = 2 W

P4.11 (b) R_{series} = 60.71 kΩ
diode $PIV \geqslant 190.9\ V$
diode current rating $\geqslant$ 1 mA average

P4.13 R = 250 Ω

P4.15 (b) $PIV \geqslant 600$ V
$I_{AV} \geqslant 12$ A

P5.1

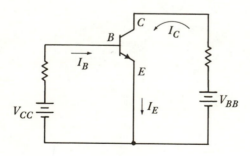

P5.3 $\beta \approx 45$

P5.5 $\alpha \approx 0.9783$

P5.7 $\beta \approx 20$

P5.9 Load line intersects axes at
20 V, 0 mA and 0 V, 10 mA

P5.11 Load line intersects axes at
30 V, 0 mA and 0 V, 20 mA

P5.13 The dc load line intersects axes at
9 V, 0 mA and 0 V, 7.965 mA
The ac load line has a slope of -1.010.
Two points one the ac load line are
V_{CE} = 5.61, I_C = 3 mA and V_{CE} = 8.580, I_C = 0

P5.15 V_{CE} (peak undistorted) = 2.970 V
V_{out} (peak undistorted) = 2.310 V

P5.17 (a) $V_{CE} \approx 14.5$ V
(b) $V_{CE} \approx 17$V
(c) As close to $V_{CE} = V_{BB}$ as signal amplitude considerations will allow

P5.19 $R_B = 93.81$ kΩ

P5.21 $R_1 = 74.67$ kΩ
$R_2 = 8$ kΩ

P5.23 $R_B = 93$ kΩ

P5.25 $A_V = -160$

P5.27 $A_I = 3.774$

P5.29 $Z_{in} = 996.3$ Ω

P5.31 $Z_{in} = 63.18$ kΩ

P6.1 See text

P6.3 See text

P6.5 $g_m = 300$ μS

P6.7 $V_{DS} = V_{BB} - I_D R_D$

P6.9 Intersects axes at
15 V, 0 mA and 0 V, 3 mA

P6.11 Intersects axes at
10 V, 0 mA and 0 V, 0.5 mA

P6.13 $R_1 = \dfrac{17}{3} R_2$

P6.15 Designer's choice

P6.17 $A_V \approx -2.75$

P6.19 $A_V \approx -2.75$

P6.21 $A_V = -83.33$

P6.23 $A_P = 139,600$

P6.25 $A_I = 1448$

P6.27 $Z_{out} = 16,666$ Ω

P6.29 $Z_{out} = 28.93$ kΩ

P8.1 $Z_{in} = 122.4$ kΩ

P8.3 $Z_{in} = 6238$ Ω

P8.5 $R_1 = 228.9$ kΩ
$R_2 = 28.13$ kΩ
Nearest standard values are
$R_1 = 220$ kΩ
$R_2 = 27$ kΩ

P8.7 $A_V = -6750$ V/V

P8.9 $Z_{out} = 27$ kΩ

P8.11 $Z_{in} = 610$ kΩ

P8.13 Forward blocking voltage ≥ 250 V
Reverse blocking voltage ≥ 250 V
$I_{av} \geq 7.958$ A

P8.15 See text

P8.17 See text

P8.19 A_V(DMDE) = 20

P8.21 See text

P8.23

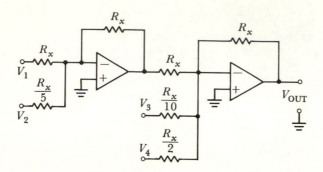

P8.25

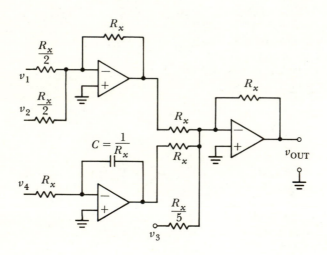

P8.27

$V_1 = 0$

$V_3 = 0$

$V_2 = -1 \text{ V}$

$V_4 = -1 \text{ V}$

$I_{RG} = 0.1 \text{ mA} \downarrow$

$I_{R1} = 0.1 \text{ mA}$

$I_{R2} = \underset{0.1 \text{ mA}}{\longrightarrow}$

$V_{R1} = -(2.5 \text{ V})+$

$V_{R2} = +(2.5 \text{ V})-$

$V_6 = -3.5 \text{ V}$

$V_5 = +2.5 \text{ V}$

$V_8 = -1.75 \text{ V}$

$V_7 = -1.75 \text{ V}$

$I_{R3} = \overline{0.425 \text{ mA}}$

$I_{R5} = \overline{0.425 \text{ mA}}$

$V_{R5} = +(4.25 \text{ V})-$

$V_9 = -6.00 \text{ V}$

P8.29 $R_G = 1020 \ \Omega$

P8.31 See text

P9.1 $A_P = 13.01$ dB

$A_P = 20$ W/W

P9.3 $A_P = 50.10$ dB

P9.5 $A_I = 3500$ A/A

$A_I = 70.88$ dB

P9.7 Designer's choice

P9.9 Designer's choice

INDEX